John Harper

Glimpses of Ocean Life

Or Rock-Pools and the Lessons they Teach

John Harper

Glimpses of Ocean Life
Or Rock-Pools and the Lessons they Teach

ISBN/EAN: 9783337034771

Printed in Europe, USA, Canada, Australia, Japan

Cover: Foto ©berggeist007 / pixelio.de

More available books at **www.hansebooks.com**

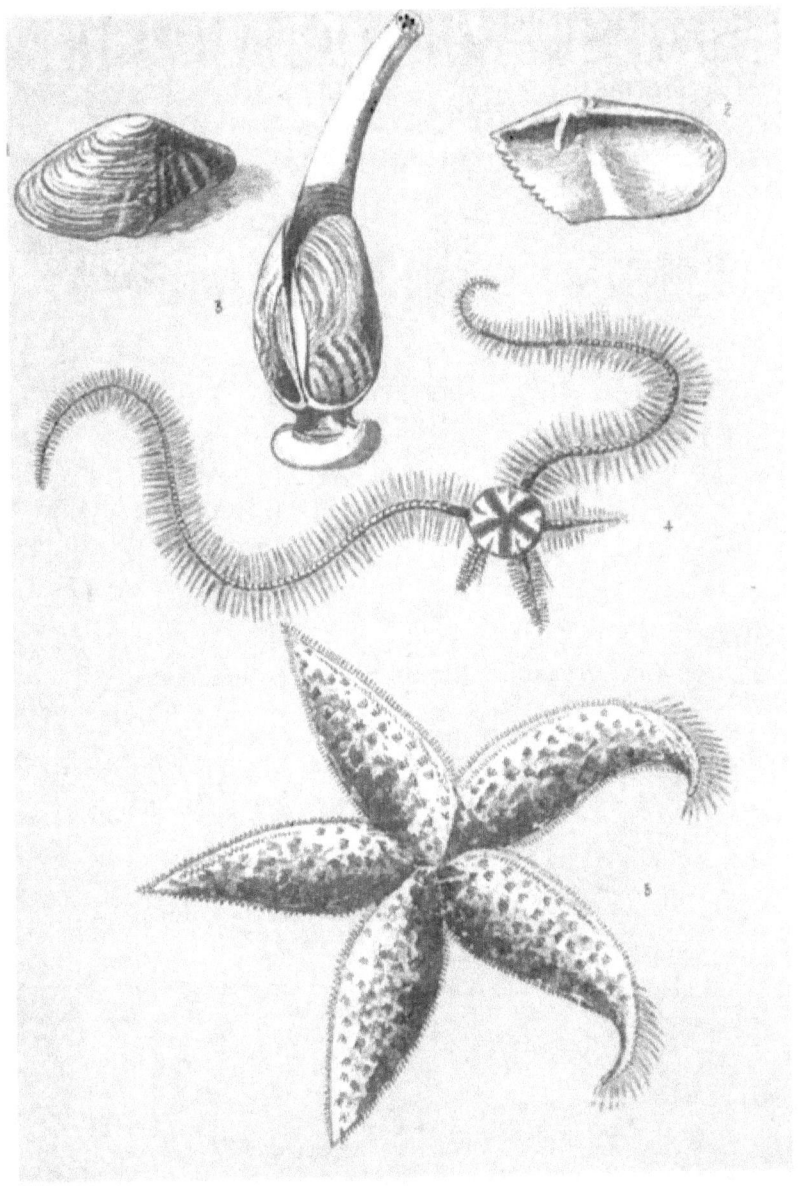

1 & 2: Valves of PHOLAS SHELL   3 *Pholas crispata*, with siphons extended
4 COMMON BRITTLE STAR *(Ophiocoma rosula)* from Nature, showing the progressive growth of new rays
5 COMMON CROSS-FISH *(Uraster rubens)*

GLIMPSES OF OCEAN LIFE;

OR,

Rock-Pools and the Lessons they Teach.

BY

JOHN HARPER, F.R.S.S.A.

AUTHOR OF 'THE SEA-SIDE AND AQUARIUM,' ETC.

WITH NUMEROUS ILLUSTRATIONS BY THE AUTHOR.

'Armado. How hast thou purchased thy experience?
Moth. By my penny of observation.'
SHAKSPEARE.

LONDON:
T. NELSON AND SONS, PATERNOSTER ROW;
EDINBURGH; AND NEW YORK.

MDCCCLXI.

TO THE RIGHT HONOURABLE

LORD BROUGHAM AND VAUX,

CHANCELLOR OF THE UNIVERSITY OF EDINBURGH,
ETC., ETC., ETC.,

THIS LITTLE VOLUME

Is Inscribed,

AS A TRIFLING TOKEN OF RESPECTFUL ADMIRATION

FOR

UNIVERSALLY RECOGNISED GREATNESS.

CONTENTS.

CHAPTER I.

ON THE PLEASURES DERIVED FROM THE STUDY OF MARINE ZOOLOGY.

Page

Introduction—Two classes of readers—Marine zoology as an amusement—The botanist and his pleasures—Entomological pursuits—Hidden marvels of nature—The little Stickleback—Conclusion, ... 17

CHAPTER II.

A GLANCE AT THE INVISIBLE WORLD.

Microscopic studies—When to use the microscope—Modern martyrs of science—Infusoria—Use of Infusoria—Distinction between plants and animals—*Vorticella*—*Rotatoria*—Wheel animalcules—Mooring Thread of Vorticellæ—A compound species of Vorticella described—*Zoothamnium spirale* of Mr. Gosse—Nature's scavengers, ... 27

CHAPTER III.

SEA ANEMONES.

Animal-flowers—*A. mesembryanthemum*—'Granny,' Sir J. Dalyell's celebrated anemone—Original anecdote—*A. troglodytes*—How to capture actiniæ—A roving 'mess.'—An intelligent anemone—Diet of the actiniæ—Voracity of these zoophytes—Defence of certain species—Actiniæ eating crabs—Their reproductive powers—Size of the 'crass.'—The Plumose anemone—Its powers of contraction, ... 45

CHAPTER IV.

EDIBLE CRAB—SHORE CRAB—SPIDER CRAB, ETC.

The Partane—Its character defended—Crustaceous demons—The wolf and the lamb—Interesting anecdote—Reason and instinct—Anecdote of the Shore crab—'The creature's run awa''—A crustaceous performer—The Fiddler crab—A little prodigal—Singular conduct of the Shore crab—The minute Porcelain crab—*Maia squinado*—*Hyas araneus*—*Maia* and *C. mænas*—Anecdote—The common Pea crab—Pinna and Pinnotheres—The Cray fish—Masticatory organs of crabs—Fishing for crabs—Crab fishers, 63

CHAPTER V.

HERMIT CRABS.

Enthusiastic students of nature—Aristocratic Hermit crabs—Swammerdam—Hermit crab and its habits—Anecdote—The Hermit in a fright—Soldier crab and Limpet—A crustaceous Diogenes—Prometheus in the tank—The martyr Hermit crab—The author's pet Blenny—Anecdote, ... 89

CHAPTER VI.

EXUVIATION OF CRUSTACEA (THE PHENOMENA OF CRABS, ETC., CASTING THEIR SHELLS).

The Tower of London—A crustaceous armory—The author's experience on the subject—Reamur and Goldsmith—Rejected shells of crabs—Anecdote—Hint to the young aquarian—Exuviation described from personal observation in several instances—Renewal of injured limbs—Frequency of exuviation—Effect of diet on crustacea—Exuviation arrested—Exuviation of the Hermit crab—How the process is effected, ... 109

CHAPTER VII.

PRAWNS AND SHRIMPS.

Habits of the Prawn—The Common Shrimp—How to catch shrimps—Conclusion, ... 135

CHAPTER VIII.

ACORN-BARNACLES—SHIP-BARNACLES.

The Common Barnacle described—Exuviation of the *Balani*—Anecdote—The Ship Barnacle—Barnacle Geese, ... 143

CHAPTER IX.

PHYLLODOCE LAMINOSA (THE LAMINATED NEREIS).

A rainy day at the sea-shore—Laminated Nereis—Its tenacity of life—Its unsuitableness for the aquarium—How the young annelids are produced—Evidence of a French naturalist, ... 151

CHAPTER X.

THE FAN-AMPHITRITE.

Its renewal of mutilated organs—How to accommodate this annelid in the tank—The 'case' of the Fan-Amphitrite, ... 159

CHAPTER XI.

THE COMMON MUSSEL.

Dr. Johnson and Bozzy—Habits of the Mussel—Marine 'at homes'—The Purpura and its habits—Enemies of the Mussel—Anecdote—Construction of the beard (or Byssus)—Author's experience—Anecdote of the mussel—Muscular action of its foot—Threads of the beard—The bridge at Bideford—Anecdote—The Mussel tenacious of life—The beard not poisonous—M. Quatrefage—Mussel beds of Esnandes—Branchiæ of the Mussel—Food of this bivalve, ... 163

CHAPTER XII.

TEREBELLA FIGULAS (THE POTTER).

Anecdote of the Potter—Its cephalic tentacula—Construction of its tubular dwelling—*Terebella littoralis*—Curious anecdote—Branchial organs of this annelid, ... 189

CHAPTER XIII.

ACALEPHÆ (MEDUSÆ, OR JELLY-FISH).

Introduction—Jelly-fish—Whales' food—Lieutenant Maury—Appearance of the Greenland Seas—Sir Walter Scott—The girdle of Venus—The Beröe—*Pulmonigrade acalephæ*—Portuguese man-of-war—*Hydra-tuba*—Alternation of generations—Dr. Reid—*Modera-formosa*—*Cyanea capillata*—Conclusion, ... 201

CHAPTER XIV.

DORIS EOLIS, ETC.

Anecdote—Young Dorides—Doris spawn—*Nudibranchiate gasteropoda*—Dr. Darwin—Mr. Gosse—A black Doris—*Bêches de mer*—A Chinese dinner—Bird's nest soup, and Sea-slug stew, ... 221

CHAPTER XV.

THE CRAB AND THE DAINTY BEGGAR.

Anecdote—The Pholas and Shore-crab—The *hyaline stylet*—The dainty beggar—The gizzard of the Pholas—Of what use is the stylet? 233

CHAPTER XVI.

THE PHOLAS, ETC. (ROCK-BORERS).

Pholades at home—Habits of the Pholas—*P. crispata*—The pedal organ—Finny gourmands—How is the boring operation performed?—Various theories on the subject—Mr Clark, Professor Owen—The Pholas at work—The boring process described from personal observation—Author's remarks on the subject—Pholas in the tank—Conclusion, ... 241

CHAPTER XVII.

THE SEA-MOUSE.

The Sea-mouse—Bristles of the aphrodite—Its beautiful plumage (?)—Its weapons of defence—The spines described—Shape of the aphrodite, &c., ... 263

CHAPTER XVIII.

STAR-FISHES, ETC.

The Coral polypes—The Lily-stars—St. Cuthbert's beads—*Pentacrinus europæus*—Rosy feather star *Ophiuridæ*—Brittle-stars—*Ophiocoma-rosula*—British asteridæ—*Uraster rubens*—Habits of this species—Submarine Dandoes—Sir John Dalyell—Professor Jones—Star-fish feeding on the oyster—Bird's foot Sea-star—*Luidia fragillissima*—Cushion-stars—Professor Forbes, ... 269

CHAPTER XIX.

SEA-URCHINS.

Sea Urchins in the tank—Growth of the Echinus—Its hedgehog-like spines—Suckers and pores—Ambulacral tubes—Professor Agassiz—Movements of the Echinus—*Pedicellariæ*—Masticatory apparatus—Common Egg Urchin—*Echinus sphæra*—How to remove the spines—'Do you boil your sea eggs?'—The Green-pea Urchin—The Silky-spined Urchin—The Rosy-heart Urchin, ... 287

CHAPTER XX.

THE SEA-CUCUMBER.

Its unattractive appearance out of water—Trepang—Several varieties eaten by the Chinese—Common Sea Cucumber—Habits of the Holothuriæ—Their self-mutilation and renewal of lost parts, ... 30

CHAPTER XXI.

THE APLYSIA, OR SEA-HARE.

Anecdote—The Sea Hare plentiful at North Berwick—Its powers of ejecting a purple fluid at certain times—Sea Hares abhorred by the ancients—Professor Forbes—Spawn of the Aplysia, ... 307

CHAPTER XXII.

SERPULÆ AND SABELLÆ.

Tubes of the *Serpulæ* — Dr. Darwin — The harbour of Pernambuco — Its wonderful structure — Reproduction of the *Serpulæ* — *Sabellæ* — Their sandy tubes, &c. ... 313

CHAPTER XXIII.

THE SOLEN, OR RAZOR FISH.

How it burrows in the sand — How specimens are caught — *Cum grano salis* — Bamboozling the Spout Fish — Amateur naturalists, and fishermen at the sea-shore, ... 321

CHAPTER XXIV.

A GOSSIP ON FISHES — INCLUDING THE ROCKLING, SMOOTH BLENNY, GUNNEL FISH, GOBY, ETC.

Punch's address to the ocean — Old blue-jackets and the 'galyant' Nelson — The ocean and its inhabitants — Life beneath the wave — Fishes the happiest of created things — A fishy discourse by St. Antony of Padua — Traveller's ne'er do lie? — The veracious Abon-el-Cassim — Do fishes possess the sense of hearing — Author's experience — An intelligent Pike fish — Dr. Warwick — The Blenny in its native haunts — A 'Little Dombey' fish — Anecdote — The Viviparous Blenny — The Gunnel fish — Five-bearded Rockling — Two-spotted Goby — Diminutive Sucker-fish — Montagu's Sucker — The Stickleback — Its nest-building habits described — Conclusion, ... 327

CHAPTER XXV.

ON THE FORMATION OF MARINE AQUARIÆ, ETC.

Mimic oceans — Practical hints on marine aquariæ — Various tanks described — The 'gravity bubble' — Evaporated sea-water — Aquariæ in France — Sea-water a contraband article across the Channel — An aquarium on a fine summer's day — The Lettuce Ulva — Author's tank — 'Excavations on a rocky shore' — Tank 'interiors' — Various centre pieces — New siphon — Aquariæ difficult to keep in hot weather — How to remove the opacity of the tank — New scheme proposed — Conclusion, ... 353

CHAPTER I.

INTRODUCTORY.

On the Pleasures derived from the Study of Marine Zoology.

'Woe to the man—
Who studies nature with a wanton eye,
Admires the work, but slips the lesson by.'

I.

As every fresh branch of investigation in natural history has a tendency to gather around it a rapidly accumulating literature, some explanation may probably be looked for from an author who offers a new contribution to the public. And when, as in the present instance, the writer's intentions are of an humble kind, it is the more desirable that he should state his views at the outset. Nor can the force of this claim be supposed to be lessened, from the gratifying fact, that the present writer has already received a warm welcome from the public.

But, before entering upon any personal explanations, it may not be out of place, in an introductory chapter such as the present, to bring under review some of the objections which have been, and still continue to be urged against this, in common with other departments of study, which are attempted to be made popular. No branch of natural history has been subjected to more disparaging opposition, partly, it must be owned, from the misplaced enthusiasm of over zealous students, than that of marine zoology.

There are two classes of readers, different in almost all other respects, whose sympathies are united in dislike of such works as this. The one, represented by men distinguished for their powers of original research, are apt to undervalue the labours of such as are not, strictly speaking, scientific writers. There is another class who, from the prejudice of ignorance, look upon marine zoology as too trivial, from the homeliness and minuteness of its details. The wonders of astronomy, and the speculations suggested by geological studies, nay, the laws of organization as exhibited in the higher forms of animal life, are clear enough to this class of readers; but it is not easy to convince them that design can be extracted from a mussel, or that a jelly-fish exhibits a marvellous power of construction.

Now, in my belief, the opposition of the better educated of these two classes of readers is the more dangerous, as it is unquestionably the more ungenerous. If Professor Ansted, when treating of the surprising neglect of geology, could thus express himself—' How many people do we meet, otherwise well educated, who look with indifference, or even contempt on this branch of knowledge,'—how much oftener may the student of the humble theme of marine zoology bewail the systematic depreciation of persons even laying claim to general scientific acquirements. This may be illustrated by an observation, made in a northern university, by a celebrated professor of Greek to a no less celebrated professor of natural history. The latter, intently pursuing his researches into the anatomy of a Nudibranche lying before him, was startled by the sudden entrance of his brother professor, who contemptuously advised him to give up skinning slugs, and take to more manly pursuits.

There is one light in which the study of marine zoology may be regarded, without necessarily offending the susceptibilities of the learned, or exciting the sneers of the ignorant. The subject may be pursued as an amusement—a pastime, if you will; and it is in no higher character than that of a holiday caterer, that the author asks the reader's company to the sea-side. No lessons but the simplest are attempted to be conveyed in this little volume, and these in as quiet and homely a style as possible.

Even in the light of an amusement, the author has something to say in behalf of his favourite study. He believes it to be as interesting, and fully as instructive as many infinitely more popular. For example: The sportsman may love to hear the whirr of the startled pheasant, as it springs from the meadow, and seeks safety in an adjoining thicket. I am as much pleased with the rustling of a simple crab, that runs for shelter, at my approach, into a rocky crevice, or beneath a boulder, shaggy with corallines and sea-weed. He, too, while walking down some rural lane, may love to see a blackbird hastily woo the privacy of a hawthorn bush, or a frightened hare limp across his path, and strive to hide among the poppies in the corn-field; I am equally gratified with the sight of a simple razor-fish sinking into the sand, or with the flash of a silver-bodied fish darting across a rock-pool.

Nay, even the trembling lark that mounts upwards as my shadow falls upon its nest among the clover, is not a more pleasant object to my eye, than the crustaceous hermit, who rushes within his borrowed dwelling at the sound of footsteps. In fact, the latter considerably more excites my kindly sympathies, from its mysterious curse of helplessness. It cannot run from danger, but can only hide itself within its shelly burden, and trust to chance for protection.

Neither the botanist nor the florist do I envy. The latter may love to gather the 'early flowrets of the year,' or pluck an opening rose-bud, but, although very beautiful, his treasures are ephemeral compared with mine.

'Lilies that fester, smell far worse than weeds.'

But I can gather many simple ocean flowers, or weeds that—

'Look like flowers beneath the flattering brine,'

whose prettily tinted fronds will 'grow, bloom, and

luxuriate' for months upon my table. They do not want careful planting, or close attention, or even—

'Like their earthly sisters, pine for drought,'

but are strong and hardy, like the pretty wild flowers that adorn our fields and hedge-rows. In the pages of an album, I can, if so disposed, feast my eyes for years upon their graceful forms, whilst their colours will remain as bright as when first transplanted from their native haunts by the sea shore.

The entomologist delights to stroll in the forest and the field, to hear the pleasant chirp of the cricket in the bladed grass, to watch the honey people bustling down in the blue bells, or even to net the butterfly as it settles on the sweet pea-blossom, while I am content to ramble along the beach, and watch the ebb and flow of the restless sea—

'So fearful in its spleeny humours bent,
So lovely in repose—'

or search for nature's treasures among the weed-clad rocks left bare by the receding tide.

A disciple of the above mentioned branch of natural history will dilate with rapture upon the wondrous transformations which many of his favourite insects undergo. But none that he can show surpasses in grandeur and beauty the changes which are witnessed in many members of the marine animal kingdom. He points to the leaf, to the bloom upon the peach, brings his microscope and bids me peer in, and behold the mysteries of creation which his instrument unfolds. 'Look,' he says, pointing to the verdant leaf, 'at the myriads of beings that inhabit this simple object. Every atom,' he exultingly exclaims, 'is a standing miracle, and adorned with such qualities, as could not be impressed upon it by a power less than infinite!' Agreed. But has not the zoologist equal reason to be proud of his science and its hidden marvels? Can he not exhibit equal miracles of divine power?

Take, as an example, one of the monsters of the deep, the whale; and we shall find, according to several learned writers, that this animal carries on its back and in its tissues a mass of creatures so minute, that their number equals that of the entire population of the globe. A single frond of marine algæ, in size

> 'No bigger than an agate stone
> On the forefinger of an alderman,'

may contain a combination of living zoophytic beings so infinitely small, that in comparison the 'fairies' midwife' and her 'team of little atomies' appear monsters as gigantic, even as the whale or behemoth, opposed to the gnat that flutters in the brightest sunbeam.

Again: in a simple drop of sea-water, no larger than the head of a pin, the microscope will discover a million of animals. Nay, more; there are some delicate sea-shells (*foraminifera*) so minute that the point of a fine needle at one touch crushes hundreds of them.

> 'Full nature swarms with life; one wondrous mass
> Of animals, or atoms organized,
> Waiting the vital breath when Parent Heaven
> Shall bid his spirit flow.'

Lastly, How fondly some writers dwell upon the many touching instances of affection apparent in the feathered tribe, and narrate how carefully and how skilfully the little wren, for example, builds its nest, and tenderly rears its young. I have often watched the common fowl, and admired her maternal anxiety to make her outspread wings embrace the whole of her unfledged brood, and keep them warm. The cat, too, exhibits this characteristic love of offspring in a marked degree. She will run after a rude hand that grasps one of her blind kittens, and, if possible, will lift the little creature, and run away home with it in her mouth. Now, whether we look at the singular skill of the bird building its nest, the hen sitting near and protecting its brood, or the cat grasping her young in its jaws, and carrying them home in safety, we shall find that all these charming traits are wonderfully combined in one of the humblest members of the finny tribe, viz., the common stickleback,—the little creature that boys catch by thousands with a worm and a pin,—that lives equally content in the clear blue sea or the muddy fresh water pool.

The author now finds that he has been much

too prolix in these preliminary observations to leave himself space for a lengthened explanation of his reasons for again intruding upon the public. These are neither original nor profound. But he cannot help expressing an earnest hope that he may get credit from old friends, and perhaps from some new, for wishing to show that the book of nature is as open as it is varied and inexhaustible; and that, however jealously guarded are many of the great secrets of organization, a knowledge of some of the most familiar objects tends to inspire us alike with wonder and with awe.

CHAPTER II.

A Glance at the Invisible World.

'There is a great deal of pleasure in prying into this world of wonders, which Nature has laid out of sight, and seems industrious to conceal from us. . . . It seems almost impossible to talk of things so remote from common life and the ordinary notions which mankind receive from the blunt and gross organs of sense, without appearing extravagant and ridiculous.'—ADDISON.

II.

It is hardly possible to write upon marine zoology without either more or less alluding to those many objects, invisible to the naked eye, which call for the use of the microscope; and it seems equally difficult for any one who has been accustomed to this instrument to speak in sober terms of its wonderful revelations. The lines of Cowper, as the youngest student in microscopic anatomy will readily acknowledge, present no exaggerated picture of ecstasy:—

> 'I have seen a man, a worthy man,
> In happy mood conversing with a fly;
> And as he through his glass, made by himself,
> Beheld its wondrous eye and plumage fine,
> From leaping scarce he kept for perfect joy.'

It is proper, however, to notice that a serious objection has been urged against the use of the microscope by young persons, namely, the injurious effects of its habitual use upon the eyesight.

So far as my experience goes, I cannot deny that this objection is well founded. Since I have begun to use the instrument, I am obliged, if I wish to view distinctly any distant object, to distort my eyes somewhat to the shape of ill-formed button-holes puckered in the sewing. Some individuals, I am aware, foolishly affect this appearance, from the notion that it exhibits an outward and visible sign of their inward profundity of character. In my own case this result may have arisen from my having worked principally at night or in the dusk. 'As to the sight being injured by a continuous examination of minute objects,' writes Mr. Clark, a most scientific naturalist, 'I can truly say this idea is wholly without foundation, if the pursuit is properly conducted; and that, on the contrary, it is materially strengthened by the use of properly adapted glasses, even of high powers; and in proof I state, that twenty years ago I used spectacles, but the continued and daily examination of these minutiæ (*foraminifera*) has so greatly increased the power of vision, that I now read the smallest type without difficulty and without aid. The great point to be attended to is not to use a power that in the least exceeds the necessity; not to continue the exercise of vision too long, and never by artificial light; and to reserve the high powers of certain lenses and the microscope for important investigations of very moderate continuance. The observant eye seizes at a glance the intelligence required; whilst strained poring and long optical exertions are delusive and unsatisfactory, and produce those fanciful imaginations of objects which have really no existence. The proper time for research after microscopic objects is for *one* hour after breakfast, when we are in the fittest state for exertion.'

Mr. Lewes, again, speaking to the same point, viz., the eyes being injured by microscopic studies, says:—'On evidence the most conclusive I deny the accusation. My own eyes, unhappily made delicate by over-study in imprudent youth, have been employed for hours daily over the microscope without injury or fatigue. By artificial light, indeed, I find it very trying; but by daylight, which on all accounts is the best light for the work, it does not produce more fatigue than any other steadfast employment of the eye. Compared with looking at pictures, for instance, the fatigue is as nothing.'

In spite of the foregoing assertions, I feel it my duty to caution the student against excess of labour. Let him ride his hobby cautiously, instead of seeking to enrol his name among the martyrs of science, of whom the noble Geoffry St. Hilaire, M. Sauvigny, and M. Strauss Dürckheim, are noted modern examples. Each member of this celebrated trio spent the latter part of his existence in physical repose, having become totally blind from intense study over the microscope. But setting aside the evils of excess, we must bear witness to the intense delight which this pursuit affords when followed with moderation.

> "'Tis sweet to muse upon the skill displayed
> (Infinite skill!) in all that *He* has made:
> To trace in Nature's most minute design
> The signature and stamp of power divine,
> Contrivance intricate, expressed with ease,
> *Where unassisted sight no **beauty sees**.*"

As my aim is merely to give the reader a taste of the subject, and whet his appetite for its more extensive pursuit at other sources, I shall confine my remarks to a few of those creatures which are readily to be found in any well-stocked aquarium. The number of animalculæ and microscopic zoospores of plants, invisible to the naked eye, with which such a receptacle is filled, even when the water is clear as crystal, is truly marvellous. These animals mostly belong to the class *Infusoria*, so called from their being found to be invariably generated in any *infusion*, or solution of vegetable or animal matter, which has begun to decay. Now, the water in an aquarium which has been kept for any length of time necessarily becomes more or less charged with the effete matter of its inhabitants, which, if allowed to accumulate, would soon render the fluid poisonous to every living thing within it. This result is happily averted by the Infusoria, which feed upon the decaying substances in solution, while they themselves become in their turn the food of the larger animals. Indeed, they constitute almost the sole nutriment of many strong, muscular shell-fish, as pholas, mussel, cockle, &c.; and doubtless help to maintain the life of others, such as actiniæ, and even crabs, which, as is well known, live and grow without any other apparent means of sustenance. Thus the presence of Infusoria in the tank may be considered a sign of its healthy condition, although their increase to such an extent as to give a milky appearance to the water, is apt to endanger the well-being of the larger, though delicate creatures. The peculiar phenomenon alluded to arises from decaying matter, such as a dead worm or limpet, which should be sought after and removed with all possible speed. The whereabouts of such objectionable remains will be generally indicated by a dense cloud of Infusoria hovering over the spot. The milkiness, however, although it may look for the time unsightly, is ofttimes the saving of the aquarium 'stock.' When these tiny but industrious scavengers have completed their task of purification, they will cease to multiply, and mostly disappear, leaving the water clear as crystal. I believe it is the absence or deficient supply of Infusoria that sometimes so tantalizingly defeats the attempts of many persons to establish an aquarium. Pure deep-sea water, although never without them, often contains but very few, hence great caution is necessary not to overstock the tank filled with it, otherwise the animals will die rapidly, although the water itself appears beautifully transparent.

Of Infusoria there are many species. They are nearly all, at one stage or other of their existence, extremely vivacious in their movements; so much so, indeed, that it becomes a matter of difficulty to observe them closely. Some have the power of darting about with astonishing velocity, others unceasingly gyrate, or waltz around with the grace of a Cellarius; while not a few content themselves by, slug-like, dragging their slow length along. The last are frequently startled from their propriety and aplomb by the rapid evolutions of their terpshicorean neighbours. Some, again, grasping hold of an object by one of their long filaments, revolve rapidly round it, whilst others spring, leap, and perform sundry feats of acrobatism that are unmatched in dexterity by any of the larger animals.

I may here observe that the motions and general structure of many of the microscopic forms of vegetation, so much resemble those of some of the infusoria, that it has long puzzled naturalists to distinguish between them with any degree of certainty. The chief distinction appears to lie in the nature of their food. Those forms which are truly vegetable can live upon purely inorganic matter, while the animals require that which is organized. The plants also live entirely by the absorption of fluid through the exterior, while the animalculæ are capable of taking in solid particles into the interior of the body. Their mode of multiplication, and the metamorphoses they undergo, are much alike in both classes, being, during one stage of their existence, still and sometimes immovably fixed to stones, sea-weed, &c., and at another freely swimming about. Notwithstanding the similarities here stated, the appearance of certain of the species is as various as it is curious. One of the commonest species of the Infusoria (*Paramecium caudatum*) is shaped somewhat like a grain of rice, with a piece chipped out on one side, near the extremity of its body. It swims about with its unchipped extremity foremost, rotating as it goes. During the milky condition of the water (before alluded to), these creatures swarm to such a degree, that a single drop of the fluid, when placed under the microscope, appears filled with a dense cloud of dancing midges. Another (*Kerona silurus*) may be said to resemble a coffee-bean, with a host of *cilia*, or short bristles, on the flat side. These are used when swimming or running. But perhaps the most singular and beautiful of all the infusorial animalcules are the *Vorticellæ*, which resemble minute cups or flower-bells, mounted upon slender retractile thread-like stalks, by which they are moored to the surface of the weeds and stones. They are called Vorticellæ on account of the little vortices or whirlpools which they continually create in the water, by means of a fringe of very minute cilia placed round the brim of their cups. These cilia are so minute as to require a very high microscopic power to make them visible, and even then they are not easily detected, on account of their extremely rapid vibration, which never relaxes while the animal is in full vigour. On the other hand, when near death, their velocity diminishes, and ample opportunity is afforded for observing that the movements consist of a rapid bending inwards and outwards, over the edge of the cup. This is best seen in a side view. The action is repeated by each cilium in succession, with such rapidity and regularity that, when viewed from above, the fringe looks like the rim of a wheel in rapid revolution. A similar appearance, produced by the same cause, in another class of animalcula, of much more complex structure than the Vorticellæ, has procured for it the name of *Rotifera*, or wheel-bearers. The result of this combined movement of the cilia is, that a constant stream of water is drawn in towards the centre of the cup, and thrown off over the sides, when, having reached a short distance beyond the edge, it circles rapidly in a small vortex, curling downwards over the lips. These currents are rendered evident by floating particles in the water. The possession of these vibratile cilia is not peculiar to this class of animals; indeed, there is good reason to believe that there is scarcely a living creature, from the lowest animalcule, or plant germ, up to man himself, that is not provided with them in some part or other. In many of these Infusoria the cilia constitute the organs of locomotion; while in the higher forms they serve various other purposes, but chiefly that of directing the flow of the various internal fluids through their proper channels. But the peculiar and perhaps most wonderful organ of the Vorticella, is its stalk or mooring thread. This though generally of such extreme tenuity as to be almost invisible with ordinary microscopes, yet exhibits a remarkable degree of strength and muscular activity in its movements, which apparently are more voluntary than those of the cilia. Its action consists of a sudden contraction from a straight to a spiral form with the coils closely packed together, by which the head or bell is jerked down almost into contact with the foot of the stalk; after a few seconds the tension seems gradually relaxed, the coils are slowly unwound, and the stalk straightens itself out. This action takes place at irregular intervals, but it is seldom that more than a minute elapses between each contraction. It (the contraction) invariably happens when the animal is touched or alarmed, and is, consequently, very frequent when the water swarms with many other swimming animalcula. When it takes place the flower-bell generally closes up into a little round ball, which opens out again only when the stalk becomes fully extended. From this we might almost infer that some animalcule, or other morsel of food, had been seized and retained within the cup; moreover, that the contraction of the stalk assisted in securing or disposing of the prey. This, however, is uncertain.

The motions of the Vorticella do not seem much affected by the stalk losing hold of its attachment;

but the result of such an accident taking place is that the cilia cause the animal to swim through the water, trailing its thread behind it, and the contraction of the latter merely causes it to be drawn up to the head.

There are various species of Vorticellæ. That just described is the simplest, consisting merely of a hemispherical ciliated cup, attached to a single thread. It is barely visible to the naked eye. But there is a compound species which I have this year found to be extremely abundant in my aquarium,—whose occupants, both large and small, it excels in singularity and beauty. In structure it is to the simple Vorticella what a many-branched zoophyte is to an *Actinia*. My attention was first drawn to the presence of this creature by observing some pebbles and fronds of green ulva thickly coated with a fine flocculent down. On closer inspection this growth appeared to consist of a multitude of feathery plumes, about one-sixteenth of an inch in height, and individually of so fine and transparent a texture as to be scarcely discernible to the unassisted sight. On touching one with the point of a fine needle it would instantly shrink up into a small but dense mass, like a ball of white cotton—scarcely so large as a fine grain of sand. In a few seconds it would again unfold and spread itself out to its original size. By carefully detaching a specimen with the point of a needle or pen-knife, and transferring it, along with a drop of water upon a slip of glass, to the stage of the microscope, a sight was presented of great wonder and loveliness :—

> 'The more I fixed mine eye,
> Mine eye the more new wonders did espye!'

Let the reader imagine a tree with slender, gracefully curved, and tapering branches thickly studded over with delicate flower-bells in place of leaves. Let him suppose the bells to be shaped somewhat between those of the fox-glove and convolvolus, and the stem, branches, bells, and all, made of the purest crystal. Let him further conceive every component part of this singular structure to be tremulous with life-like motion, and he will have as correct an idea as words can give of the complex form of this minute inhabitant of the deep. Moreover, while gazing at it through the microscope, the observer is startled by the sudden collapse of the entire structure. The lovely tree has shrunk together into a dense ball, in which the branching stem lies completely hidden among the flower-bells—themselves closed up into little spherules, so closely packed together that the entire mass resembles a piece of herring-roe. This contraction is so instantaneous that the mode in which it is accomplished cannot be observed until the tree is again extended. As the re-extension takes place very slowly, we are enabled to observe that each branchlet has been coiled in a spiral form, like the thread of the simple Vorticella previously described; and also that the main stem, above the lowest branch, was coiled up in the same way, but not so closely, and that the part below the lowest branch had, curiously enough, remained straight. Sometimes, in large and numerously branched specimens, one or two of the lowest members do not contract at the same time with the rest, but do so immediately afterwards, as if they had been startled by the shrinking movements of their neighbours. Sometimes these lowest branches will contract alone, while all the others remain fully extended,—a fact that would almost seem to indicate that they possessed an independent life of their own.

In the accompanying engraving I have attempted faithfully to portray one of these wonderful creatures. Fig. 1 represents it fully extended, while Fig. 2 indicates its collapsed form. There is another curious circumstance which I have fortunately observed in connection with this Vorticella, a description of which will perhaps be interesting to the reader. I allude to the casting off of what may be called the fruit of the tree. When this event takes place, the buds (or fruit) dart about with such rapidity, that it is almost impossible to keep them in the field of view for the briefest space of time. A represents the enchanted fruit hanging on the tree; B shows it as it swims about.

Although not exactly fruit, it is, no doubt, the means by which the Vorticellæ are propagated, for it is known that many fixed zoophytes, and even some plants, produce free swimming germs or spores, which

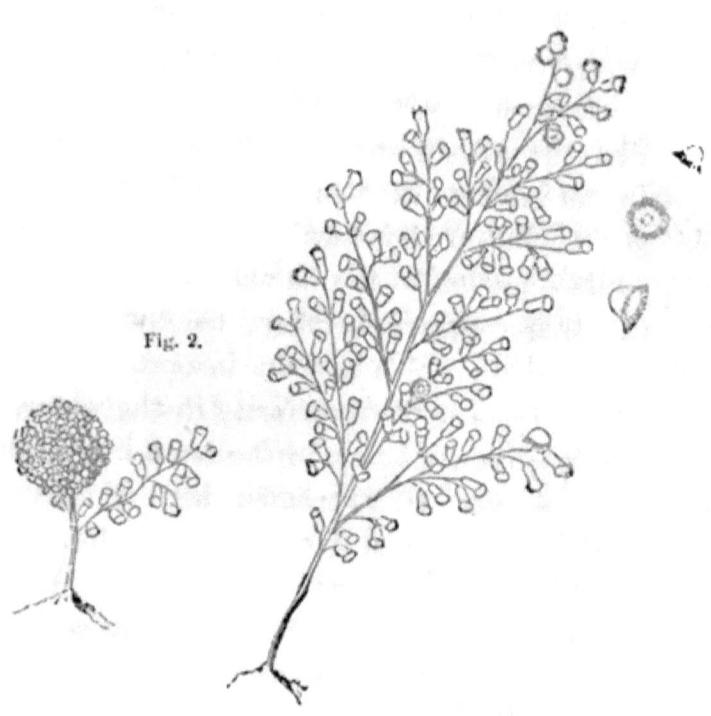

Fig. 1.

Fig. 2.

afterwards become fixed, and grow up into forms like those which produced them. In some of the branching zoophytes (*Coryne, Sertularia,* &c.), the germs are exactly like little medusae, being small, gelatinous cups fringed with tentacula, by means of which they twitch themselves along with surprising agility. In this Vorticella, however, it is more like one of the ciliated Infusoria. The first one that I saw attached I conceived to be a remarkably large bell, with its mouth directed towards me, but the cilia with which it appeared to be fringed were unusually large and distinct. The movements of these appendages being comparatively slow, it was most interesting to watch them as they successively bent inwards and rose again, like the steady swell of a tidal wave, or an eccentric movement in some piece of machinery, making a revolution about twice in a second, and in the opposite direction to the hands of a clock. Suddenly the tree contracted, when, to my surprise, I observed the bell, which not an instant before appeared attached, now floating freely in the water, its ciliary movements not being in the least interrupted. Presently, however, they became brisker, the bell turned over on its side, and, ere the tree had again expanded, darted out of view, not, however, before I had remarked that it was not a bell, but a sphere flattened on one side, and having its circular ring of cilia on the flat side, with only a slight depression in the middle of it. There also appeared to be a small granular nucleus immediately above this depression, the rest of the body being perfectly transparent. I afterwards saw several others attached to the tree, each seated about the centre of a branch; but none of these were so fully developed. They were like little transparent button mushrooms, and had all more or less of a nucleus on the side by which they were attached. On only one of these did I detect any cilia.

Mr. Gosse, in his 'Tenby,' gives a picture of an animal exceedingly like what I have described; but from his account of it, there seems to be some doubt of their identity. He calls it '*Zoothamnium spirale*,' because the insertions of the branches were placed spirally around the main stem, like those of a fir-tree. In my specimens the branches were set alternately on opposite sides of the main trunk, and the whole was curved like a drooping fern leaf or an ostrich feather, the bells being mostly set on the convex side.

In conclusion, let me mention that it is an error to suppose, as many persons do, that putrid water alone contains life. Infusoria occur, as before hinted, in the clear waters of the ocean, in the water that we drink daily, and also in the limpid burn that flows through our valleys, or trickles like a silver thread down the mountain side.*

> Where the pool
> Stands mantled o'er with green, invisible,
> Amid the floating verdure millions stray.
> Each liquid too, whether it pierces, soothes,
> Inflames, refreshes, or exalts the taste,
> With various forms abounds. Nor is the stream
> Of purest crystal, nor the lucid air,
> Though one transparent vacancy it seems,
> Void of their unseen people. These, concealed
> By the kind art of forming Heaven, escape
> The grosser eye of man.'

---

* Ehrenberg states that Infusoria are in a higher state of organization when taken from pure streams than from putrid waters.

Let it be remembered, too, that Infusoria, when found in either do not themselves constitute the impurity of fresh or salt water; they merely act as 'nature's invisible scavengers,' whose duty it is to remove all nuisances that may spring up; and most unceasingly do these tiny creatures labour in the performance of their all-important mission of usefulness.

CHAPTER III.

Sea Anemones.

'The living flower that, rooted to the rock,
　Late from the thinner element,
　Shrunk down within its purple stem to sleep,
　Now feels the water, and again
　Awakening, blossoms out
　All its green anther-necks.'

1 Sir J.G Dalyell's celebrated ACTINIA. (Drawn from Nature Jan. 1860.)
2 A CRASSICORNIS
3 CAVE DWELLER *(A troglodytes.)*

III.

No marine objects have become more universally popular of late years than Sea Anemones. Certainly none better deserve the attention which has been, and is daily bestowed upon them by thousands of amateur naturalists, who cannot but be delighted with the wondrous variety of form, and the beauteous colouring which these zoophytes possess.

A stranger could scarcely believe, on looking into an aquarium, that the lovely object before him, seated motionless at the base of the vessel, with tentacula expanded in all directions, was not a simple daisy newly plucked from the mountain side, or it may be a blooming marigold or *Anemone* from some rich parterre—instead of being, in reality, a living, moving, animal-flower.

One great advantage which the *Actiniæ* possess over certain other inhabitants of the sea shore, at least to the eye of the naturalist, is the facility with which specimens may be procured for observation and study. Scarcely any rock-pool near low water mark but will be found to encompass a certain number of these curious creatures, while some rocky excavations of moderate size will at times contain as many as fifty. Should the tide be far advanced, the young zoologist need not despair of success, for, by carefully examining the under part of the boulders totally uncovered by the sea, he will frequently find specimens of the smooth anemone, contracted and hanging listlessly from the surface of the stone, like masses of green, marone, or crimson jelly.

The Actiniæ, and especially examples of the above mentioned species, are extremely hardy and tenacious of life, as the following interesting narrative will prove.

The late Sir John Dalyell writing in 1851, says, ' I took a specimen of *A. mesembryanthemum* (smooth anemone) in August 1828, at North Berwick, where the species is very abundant among the crevices of the rocks, and in the pools remaining still replenished after the recess of the tide. It was originally very fine, though not of the largest size, and I computed from comparison with those bred in my possession, that it must have been then at least seven years old.'

Through the kindness of Dr. M'Bain, R.N., the writer has been permitted to enjoy the extreme pleasure of inspecting the venerable zoophyte above alluded to, which cannot now be much under thirty-eight years of age !

In the studio of the above accomplished naturalist, 'Granny' (as she has been amusingly christened) still dwells, her wants being attended to with all that tenderness and care which her great age demands.

Sir J. Dalyell informs us that during a period of twenty years this creature produced no less than 344 young ones. But, strange to say, nearly the fortieth part of this large progeny consisted of monstrous animals, the monstrosity being rather by redundance than defect. One, for instance, was distinguished by two mouths of unequal dimensions in the same disc, environed by a profusion of tentacula. Each mouth fed independently of its fellow, and the whole system seemed to derive benefit from the repast of either. In three years this monster became a fine specimen, its numerous tentacula were disposed in four rows, whereas only three characterize the species, and the tubercles of vivid purple, regular and prominent, at that time amounted to twenty-eight.

From the foregoing statement we learn that this extraordinary animal produced about 300 young during a period of twenty years, but, 'wonder of wonders!' I have now to publish the still more surprising fact, that in the spring of the year 1857, after being unproductive for many years, it unexpectedly gave birth, during a single night, to no less than 240 living models of its illustrious self!

This circumstance excited the greatest surprise and pleasure in the mind of the late Professor Fleming, in whose possession this famous Actinia then was.

Up to this date (January 1860) there has been no fresh instance of fertility on the part of Granny, whose health, notwithstanding her great reproductive labours and advanced age, appears to be all that her warmest friends and admirers could desire. Nor does her digestive powers exhibit any signs of weakness or decay; on the contrary, that her appetite is still exquisitely keen, I had ample opportunity of judging. The half of a newly opened mussel being laid gently upon the outer row of tentacula, these organs were rapidly set in motion, and the devoted mollusc engulphed in the course of a few seconds.

The colour of this interesting pet is pale brown. Its size, when fully expanded, no larger than a half-crown piece. It is not allowed to suffer any annoyance by being placed in companionship with the usual occupants of an aquarium, but dwells alone in a small tank, the water of which is changed regularly once a week. This being the plan adopted by the original owner of Granny, is the one still followed by Dr. M'Bain, whose anxiety is too great to allow him to pursue any other course, for fear of accident thereby occurring to his protegée.

A portrait of Granny, drawn from nature, will be found on Plate 2.

*A. troglodytes*** (cave-dweller) is a very common, but interesting object. The members of this species are especial favourites with the writer, from their great suitableness for the aquarium. They vary considerably in their appearance from each other. Some are red, violet, purple, or fawn colour; others exhibit a mixture of these tints, while not a few are almost entirely white. There are certain specimens which disclose tentacula, that in colour and character look, at a little distance, like a mass of eider-down spread out in a circular form. A better comparison, perhaps, presents itself in the smallest plumage of a bird beautifully stippled, and radiating from a centre. The centre is the mouth of the zoophyte, and is generally a light buff or yellow colour. From each corner, in certain specimens, there branches out a white horn that tapers to a very delicate point, and is ofttimes gracefully curled like an Ionic volute, or rather like the tendril of a vine.

In addition to the pair of horns alluded to, may sometimes be seen a series of light-coloured rays, occurring at regular intervals around the circumference of the deep tinted tentacula, and thereby produc-

---

* The above mentioned Actinia is extremely abundant on the shores of the Frith of Forth. Sir J. Dalyell terms it *A. explorator*. Local amateur naturalists frequently reject the specific name of 'Troglodytes,' and adopt the more musical appellation of 'Daisy-Anemone.' Such error seems very pardonable, when we remember the close resemblance which the creature when expanded bears to the daisy of the field. In no single instance have I met with specimens of the true *A. bellis* at the above named locality, nor do I think any have ever been found by previous naturalists.

ing to the eye of the beholder a most pleasing effect.

As a general rule, never attempt to capture an anemone unless it be fully expanded, before commencing operations. By this means you will be able to form a pretty accurate estimate of its appearance in the tanks. This condition of being seen necessitates, of course, its being covered with water, and, consequently, increases the difficulty of capturing your prize, especially when the creature happens to have taken up a position upon a combination of stone and solid rock, or in a crevice, or in a muddy pool, which when disturbed seems as if it would never come clear again.

It is, in consequence, advisable to search for those situated in shallow water, the bottom of which is covered with clean sand. When such a favourable spot is found, take hammer and chisel and commence operations. Several strokes may be given before any alarm is caused to the anemone, provided it be not actually touched. No sooner, however, does the creature feel a palpable vibration, and suspect the object of such disturbance, than, spurting up a stream of water, it infolds its blossom, and shrinks to its smallest possible compass. At same time apparently tightens its hold of the rock, and is, indeed, often enabled successfully to defy the utmost efforts to dislodge it.

After a little experience, the zoologist will be able

to guess whether he is likely to succeed in getting his prize perfect and entire; if not, let me beg of him not to persevere, but immediately try some other place, and hope for better fortune.

Although apparently sedentary creatures, the Actiniæ often prove themselves to be capable of moving about at will over any portion of their subaqueous domain. Having selected a particular spot, they will ofttimes remain stationary there many consecutive months. A smooth anemone that had been domesticated for a whole year in my aquarium thought fit to change its station and adopt a roving life, but at last 'settled down,' much to my surprise, upon a large mussel suspended from the surface of the glass. Across both valves of the mytilus the 'mess.,' attached by its fleshy disc, remained seated for a considerable length of time. It was my opinion that the mussel would eventually be sacrificed. Such, however, was not the case, for on the zoophyte again starting off on a new journey, the mollusc showed no palpable signs of having suffered from the confinement to which it had so unceremoniously been subjected.

The appearance of this anemone situated several inches from the base of the vessel, branching out from such an unusual resting-place, and being swayed to and fro, as it frequently was, by the contact of a passing fish, afforded a most pleasing sight to my eye. Indeed, it was considered for a while one of the 'lions' of the tank, and often became an object of admiration not only to my juvenile visitors, but also to many 'children of larger growth.'

There is a curious fact in connection with the Actiniæ which deserves to be chronicled here. I allude to the apparent instinct which they possess. This power I have seen exercised at various times. The following is a somewhat remarkable instance of the peculiarity in question.

In a small glass vase was deposited a choice *A. dianthus*, about an inch in diameter. The water in the vessel was at least five inches in depth. Having several specimens of the *Aplysiæ*, I placed one in companionship with the anemone, and was often amused to observe the former floating head downward upon the surface of the water. After a while it took up a position at the base of the vase, and remained there for nearly a week. Knowing the natural sluggishness of the animal, its passiveness did not cause me any anxiety. I was rather annoyed, however, at observing that the fluid was becoming somewhat opaque, and that the Dianthus remained entirely closed, and intended to find out the cause of the phenomena, but from some reason or other failed to carry out this laudable purpose at the time. After the lapse of a few days, on looking into the tank, I was delighted to perceive the lace-like tentacula of the actinia spread out on the surface of the water, which had become more muddy-looking than before.

I soon discovered that the impurity in question

arose from the Aplysia (whose presence in the tank I had forgotten) having died, and its body being allowed to remain in the vessel in a decaying state. The deceased animal on being removed emitted an effluvium so intolerably bad that it seemed like the concentrated essence of vile odours. The water, of course, must have been of the most deadly character, yet had this most delicate of sea anemones existed in it for several consecutive days.

In order further to test how long my little captive would remain alive in its uncongenial habitation, I cruelly refused to grant any succour, but must own to having felt extremely gratified at perceiving, in the course of a few days, that instead of remaining with its body elongated to such an unusual extent, the Dianthus gradually advanced along the base, then up the side of the vessel, and finally located itself in a certain spot, from which it could gain easy access to the outer atmosphere.

After this second instance of intelligence (?) I speedily transferred my pet to a more healthy situation.

Having procured a small colony of Actiniæ, you need be under no anxiety about their diet, for they will exist for years without any further subsistence than is derived from the fluid in which they live. Yet strange as the statement will appear to many persons, the Actiniæ are generally branded with the character of being extremely greedy and voracious.

'Nothing,' says Professor Jones, 'can escape their deadly touch. Every animated thing that comes in contact with them is instantly caught, retained, and mercilessly devoured. Neither strength nor size, nor the resistance of the victim, can daunt the ravenous captor. It will readily grasp an animal, which, if endowed with similar strength, advantage, and resolution, could certainly rend its body asunder. It will endeavour to gorge itself with thrice the quantity of food that its most capacious stomach is capable of receiving. Nothing is refused, provided it be of animal substance. All the varieties of the smaller fishes, the fiercest of the crustacea, the most active of the annelidans, and the soft tenants of shells among the mollusca, all fall a prey to the Actiniæ.'

This is a sweeping statement, and, although corroborated by Sir J. Dalyell and others, is one that requires to be received with a certain degree of caution. It most certainly does not apply to *A. bellis, A. parisitica, A. dianthus, troglodytes*, or any other members of this group; and to a very limited extent only is it applicable to *A. coriacea* or *A. mesembryanthemum*.

As may readily be conceived, the writer could not keep monster specimens, such as are often found at the sea shore; but surely if the statement were correct that, *as a general rule*, the actiniæ eat living crabs, the phenomenon would occasionally occur with moderate-sized specimens, when kept in companionship with a mixed assembly of crustaceans. Yet in no single instance have I witnessed a small crab sacrificed to the gluttony of a small anemone.

With regard to *A. mesembryanthemum, A. bellis,* and *A. dianthus,* they get so accustomed to the presence of their crusty neighbours, as not to retract their expanded tentacula when a hermit crab, for instance, drags his lumbering mansion across, or a fiddler crab steps through the delicate rays, like a sky terrier prancing over a bed of tulips.

Thus much I have felt myself called upon to say in defence of certain species of Actiniæ; but with regard to *A. crassicornis,* I must candidly own the creature is greedy and voracious to an extreme degree.

Like many other writers, I have seen scores of this species of Actiniæ that contained the remains of crabs of large dimensions, but at one time considered that the latter were dead specimens, which had been drifted by the tide within reach of the Actiniæ, and afterwards consumed. That such, indeed, was the correct explanation in many instances I can scarcely doubt, from the disproportionate bigness of the crabs as compared with the anemones, but feel quite confident, that in other instances, the crustacea were alive when first caught by their voracious companions.

To test the power of the 'crass.,' I have frequently chosen a specimen well situated for observation, and dropped a crab upon its tentacula. Instantly the intruding animal was grasped (perhaps merely by a claw), **but in spite of** its struggles to escape, was slowly drawn into the mouth of its captor, and eventually consumed. In one **case, after** the crab had been lost to view for the space of three minutes only, I drew it out of the Actinia, but although **not** quite dead, it evidently did not seem likely to survive for any length of time.

In collecting Actiniæ great care should be taken in detaching them from their position. If possible, it is far the better plan not to disturb them, but to transport them to the aquarium on the piece of rock or other substance to which they may happen to be affixed. **This can in general be done** by a smart blow of the chisel and hammer.

Should the attempt fail, an endeavour should be made to insinuate the finger nails under the base, and so detach each specimen uninjured. This operation is a delicate one, requiring practice, **much** patience, and no little skill. We are told by some authors that a slight rent is of no consequence, since the anemone is represented as having **the power of** darning it up. It **may be so**, but for my part I am inclined in other instances to consider the statement more facetious than truthful. In making this remark, I allude solely to the disc of the animal, an injury to which I have never seen repaired. On the other hand, it is well known that certain other parts may be destroyed with impunity. If the tentacula, for instance, be cut away, so great are the reproductive powers of the Actiniæ, that in a comparatively short space of time the mutilated members will begin to bud anew.

'If cut transversely through the middle, the lower portion of the body will after a time produce more tentacula, pretty near as they were before the operation, while the upper portion swallows food as if nothing had happened, permitting it indeed at first to come out at the opposite end; just as if a man's head being cut off would let out at the neck the bit taken in at the mouth, but which it soon learns to retain and digest in a proper manner.'

The smooth anemone being viviparous, as already hinted, it is no uncommon circumstance for the naturalist to find himself unexpectedly in possession of a large brood of infant zoophytes, which have been ejected from the mouth of the parent.

There is often an unpleasant-looking film surrounding the body of the Actiniæ. This 'film' is the skin of the animal, and is cast off very frequently. It should be brushed away by aid of a camel hair pencil. Should any rejected food be attached to the lips, it may be removed by the same means. When in its native haunts this process is performed daily and hourly by the action of the waves. Such attention to the wants of his little captives should not be grudgingly, but lovingly performed by the student. His labour frequently meets with ample reward, in the improved appearance which his specimens exhibit. Instead of looking sickly and weak, with mouth pouting, and tentacula withdrawn, each little pet elevates its body and gracefully spreads out its many rays, apparently for no other purpose than to please its master's eye.

A. mesembryanthemum (in colloquial parlance abbreviated to 'mess.'), is very common at the seashore. It is easily recognised by the row of blue torquoise-like beads, about the size of a large pin's head, that are situated around the base of the tentacula. This test is an unerring one, and can easily be put in practice by the assistance of a small piece of stick, with which to brush aside the overhanging rays.

A. crassicornis grows to a very large size. Some specimens would, when expanded, cover the crown of a man's hat, while others are no larger than a 'bachelor's button.' Unless rarely marked, I do not now introduce the 'crass.' into my tanks, from a dislike, which I cannot conquer, to the strange peculiarity which members of this species possess, of turning themselves inside out, and going through a long series of inelegant contortions. Still, to the young zoologist, this habit will doubtless be interesting to witness. One author has named these large anemones 'quilled dahlias;' and the expression is so felicitous, that if a stranger at the sea-side bear it in mind, he could hardly fail to identify the 'crass,' were he to meet with a specimen in a rocky pool. Not the least remarkable feature in connection with these animal-flowers, is the extraordinary variety of colouring which various specimens display.

A. troglodytes, is seldom found larger than a florin. Its general size is that of a shilling. From the description previously given, the reader will be able to make the acquaintance of this anemone without any trouble whatever.

A. dianthus (Plumose anemone), is one of the most delicately beautiful of all the Actiniæ; it can, moreover, be very readily identified in its native haunts. Its colour is milky-white,—body, base, and tentacula, all present the same chaste hue. Specimens, however, are sometimes found lemon-coloured, and occasionally of a deep orange tint. Various are the forms which this zoophyte assumes, yet each one is graceful and elegant.

The most remarkable as well as the most common shape, according to my experience, is that of a lady's corset, such as may often be seen displayed in fashionable milliners' windows. Even to the slender waist, the interior filled with a mass of lace-work, the rib-like streaks, and the general contour, suggestive of the Hogarthian line of beauty, the likeness is sustained.

When entirely closed, this anemone, unlike

many others, is extremely flat, being scarcely more than a quarter of an inch in thickness; indeed, so extraordinary is the peculiarity to which I allude, that a novice would have great difficulty in believing that the object before him was possessed of expansive powers at all, whereas, in point of fact, it is even more highly gifted in this respect than any other species of Actiniæ.

CHAPTER IV.

Edible Crab, Shore-Crab, Spider-Crab, &c.

'With a smart rattle, something fell from the bed to the floor; and disentangling itself from the death drapery, displayed a large pound *Crab*. . . . Creel Katie made a dexterous snatch at a hind claw, and, before the Crab was at all aware, deposited him in her patch-work apron, with a "*Hech, sirs, what for are ye gaun to let gang siccan a braw partane?*"'—T. HOOD

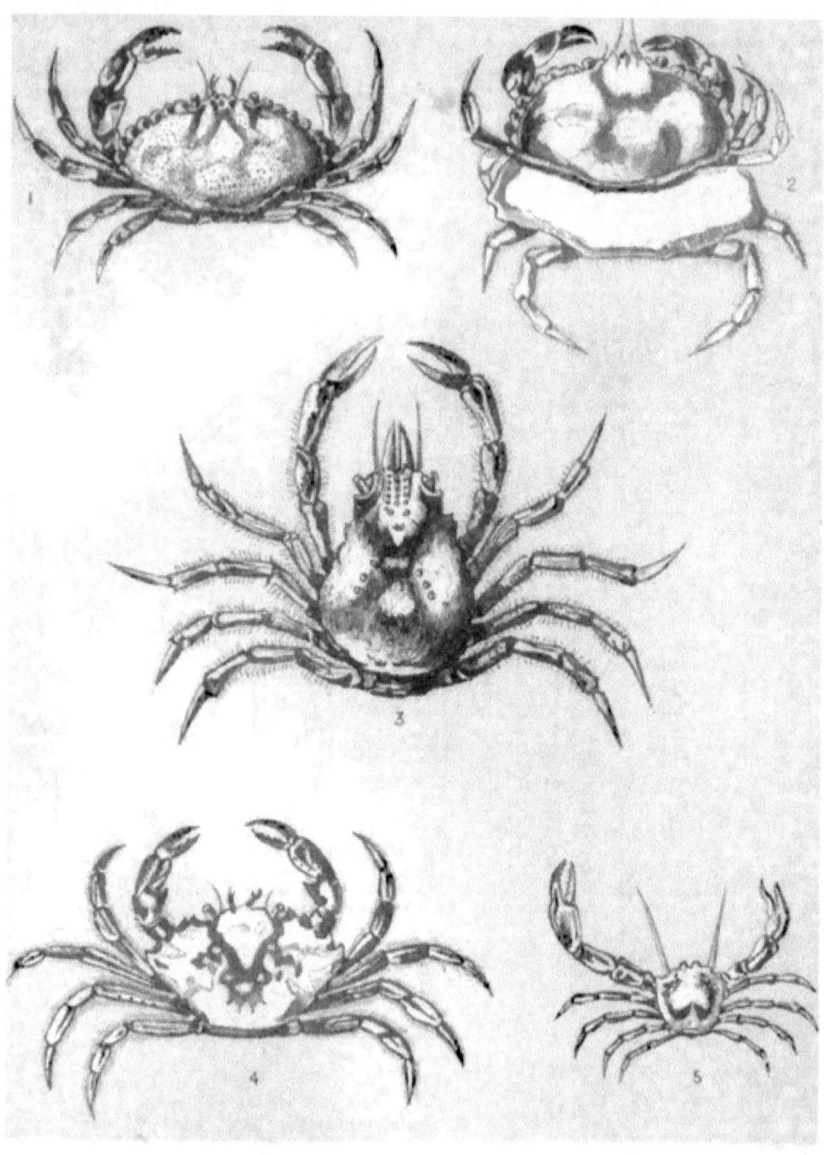

1 EDIBLE CRAB
2 EDIBLE CRAB, casting its shell, (from Nature)
3 SPIDER CRAB
4 COMMON SHORE-CRAB
5 MINUTE PORCELAIN-CRAB

IV.

The foregoing motto, extracted from a humorous tale by 'dear Tom Hood,' which appeared in one of his comic annuals,—or volumes of 'Laughter from year to year,' as he delighted to call them,—may not inaptly introduce the subject of this chapter.

The term *partane* is generally applied in Scotland to all the true crabs (*Brachyura*). An esteemed friend, however, informs me that in some parts it is more particularly used to denote the Edible Crab (*Cancer pagurus*), which is sold so extensively in the fishmongers' shops. However that may be, there is no doubt it was a specimen of this genus that Creel Katie so boldly captured.

Now this crab, to my mind, is one of the most interesting objects of the marine animal kingdom, and I would strongly advise those of my readers who may have opportunities of being at the sea-side to procure a few youthful specimens. Its habits, according to my experience, are quite different from those of its relative, the Common Shore-Crab (*Carcinus mænas*), or even the Velvet Swimming-Crab (*Portunus puber*). Unlike these, it does not show any signs of a vicious temper upon being handled, nor does it scamper away in hot haste at the approach of a stranger. Its nature, strange as the statement may appear to many persons, seems timid, gentle, and fawn-like.

On turning over a stone, you will perhaps perceive, as I have often done, three or four specimens, and, unless previously aware of the peculiarity of their disposition, you will be surprised to see each little fellow immediately fall upon his back, turn up the whites of his eyes, and bring his arms or claws together,—

> 'As if praying dumbly,
> Over his breast:'

making just such a silent appeal for mercy as a pet spaniel does when expecting from his master chastisement for some *faux pas*. One of these crabs may be taken up and placed in the hand without the slightest fear. It will not attempt to escape, but will passively submit to be rolled about, and closely examined at pleasure. Even when again placed in its native element, minutes will sometimes elapse before the little creature can muster up courage to show his 'peepers,' and gradually unroll its body and limbs from their painful contraction.

Most writers on natural history entertain an opinion totally at variance with my own in regard to the poor *Cancer pagurus*, of whom we are speaking.

By some he is called a fierce, cannibalistic, and remorseless villain, totally unfit to be received into respectable marine society. Mr. Jones relates how he put half a dozen specimens into a vase, and on the following day found that, with the exception of two, all had been killed and devoured by their companions; and in a trial of strength which speedily ensued between the pair of 'demons in crustaceous guise,' one of these was eventually immolated and devoured by his inveterate antagonist. Sir J. Dalyell mentions several similar instances of rapacity among these animals. Now, these anecdotes I do not doubt, but feel inclined, from the results of my own experience, to consider them exceptional cases.

When studying the subject of exuviation, I was in the habit of keeping half a dozen or more specimens of the Edible Crab together as companions in the same vase; but except when a 'friend and brother' slipped off his shelly coat, and thus offered a temptation too great for crustaceous nature to withstand, I do not remember a single instance of cannibalism. True, there certainly were occasionally quarrelling and fighting, and serious nocturnal broils, whereby life and limb were endangered; but then such mishaps will frequently occur, even in the best regulated families of the higher animals, without these being denounced as a parcel of savages.

Compared to *Cancer pagurus*, the Shore-Crab appears in a very unamiable light. When the two are kept in the same vase, they exhibit a true exemplification of the wolf and the lamb. This, much to my chagrin, was frequently made evident to me, but more particularly so on one occasion, when I was, from certain circumstances, compelled to place a specimen of each in unhappy companionship. Here is a brief account of how they behaved to each other: The poor little lamb (*C. pagurus*) was kept in a constant state of alarm by the attacks of her fellow-prisoner (*C. mænas*) from the first moment that I dropped her in the tank. If I gave her any food, and did not watch hard by until it was consumed, the whole meal would to a certainty be snatched away. Not content with his booty, the crabbie rascal of the shore would inflict a severe chastisement upon his rival in my favour, and not unfrequently attempt to wrench off an arm or a leg out of sheer wantonness. To end such a deplorable state of matters, I very unceremoniously took up wolf, and lopped off one of his large claws, and also one of his hind legs. By this means I stopped his rapid movements to and fro, and, moreover, deprived him somewhat of his power to grasp an object forcibly. In spite of his mutilations, he still exhibited the same antipathy to his companion, and, as far as possible, made her feel the weight of his jealous ire. Retributive justice, however, was hanging over his crustaceous head. The period arrived when nature compelled him to change his coat. In due time the mysterious operation was performed, and he stood forth a new creature, larger in size, handsomer in appearance, but for a few days weak, sickly, and defenceless. His back, legs, and every part of his body were of the consistency of bakers' dough. The lamb well knew her power, and though much smaller in size than her old enemy, she plucked up spirit and attacked him; nor did she desist until she had seemingly made him cry peccavi, and run for his life beneath the shelter of some friendly rock. Without wishing to pun, I may truly say the little partane came off with *eclat*, having my warmest approbation for her conduct, and a *claw* in her arms as token of her prowess. I knew that when wolf was himself again there would be a scene. Reprisals, of course, would follow. Therefore, rather than permit a continuance of such encounters, I separated the crabs, and introduced them to companions more suited to the nature of each.

The difference exhibited in the form and development of the tail in the ten-footed crustacea (*Decapoda*)—as for instance, the crab, the lobster, and the hermit-crab—is so striking that naturalists have very appropriately divided them into three sections, distinguished by terms expressive of these peculiarities of structure: 1st, *Brachyura*, or short-tailed decapods, as the Crabs; 2d, *Anomoura*, or irregular tailed, as the Hermit-crabs; 3d, *Macroura*, or long-tailed, as Lobster, Cray-fish, &c.

It is to a further consideration of a few familiar examples of the first mentioned group that I propose to devote the remainder of this chapter.

Few subjects of study are more difficult and obscure than such as belong to the lower forms of the animal kingdom. However carefully we may observe the habits of these animals, our conclusions are too often apt to be unsound, from our proneness to judge of their actions as we would of the actions of men. As a consequence, an animal may be pronounced at one moment quiet and intelligent, and at another obstinate and dull, while perhaps, if the truth were known, it deserves neither verdict.

For my own part, the more I contemplate the habits of many members of the marine animal kingdom, the more am I astounded at the seeming intelligence and purpose manifested in many of their actions. Prior, apparently, must have been impressed with the same idea, for he says, speaking of animals,—

> 'Vainly the philosopher avers
> That reason guides our deeds, and instinct *theirs*.
> How can we justly different causes frame
> *When the effects entirely are the same?*
> Instinct and reason, how can we divide?
> 'Tis the fool's ignorance, and the pedant's pride!"

This train of thought has been suggested to my mind by viewing the singular conduct of a Shore-Crab, whom I kept domesticated for many consecutive months. Three times during his confinement he cast his exuvium, and had become nearly double his original size. His increased bulk made him rather unfit for my small ocean in miniature, and gave him, as it were, a loblolliboy appearance. Besides, he was always full of mischief, and exhibited such pawkiness, that I often wished he were back again to his sea-side home. Whenever I dropped in a meal for my Blennies, he would wait until I had retired, and then rush out, disperse the fishes, and appropriate the booty to himself. If at all possible, he would catch one of my finny pets in his arms, and speedily devour it. Several times he succeeded in so doing; and fearing that the whole pack would speedily disappear, unless stringent measures for their preservation were adopted, I determined to eject the offender. After considerable trouble, his crabship was captured, and transferred to a capacious glass.

The new lodging, though not so large as the one to which for so long a time he had been accustomed, was nevertheless clean, neat, and well-aired. At its base stood a fine piece of polished granite, to serve as a chair of state, beneath which was spread a carpet of rich green ulva. The water was clear as crystal; in fact, the accommodation, as a whole, was unexceptionable. The part of host I played myself, permitting no one to usurp my prerogative. But in spite of this, the crab from the first was extremely dissatisfied and unhappy with the change, and for hours together, day after day, he would make frantic and ineffectual attempts to climb up the smooth walls of his dwelling-place. Twice a day, for a week, I dropped in his food, consisting of half a mussel, and left it under his very eyes; nay, I often lifted him up and placed him upon the shell which contained his once-loved meal; still, although the latter presented a most inviting come-and-eat kind of appearance, not one particle would he take, but constantly preferred to raise himself as high as possible up the sides of the vase, until losing his balance, he as constantly toppled over and fell upon its base.

This behaviour not a little surprised me. Did it indicate sullenness? or was it caused by disappointment? Was he aware that escape from his prison without aid was impossible, and consequently exhibited the pantomime, which I have described, to express his annoyance, and longing for the home he had lately left?

Thinking that perhaps there was not sufficient sea-weed in the glass, I added a small bunch of *I. edulis*. Having thus contributed, as I believed, to the comfort of the unhappy crab, I silently bade him *bon soir*. On my return home, I was astonished by the servant, who responded to my summons at the door, blurting out in a nervous manner, 'O sir! the creature's run awa!' 'The creature—what creature?' I inquired. 'Do ye no ken, sir?—the wee crabbie in the tumler!'

I could scarcely credit the evidence of my sight

when I saw the 'tumler' minus its crustaceous occupant. The first thought that occurred to me was as to where the crab could be found. Under chairs, sofa, and fender, behind book-case, cabinet, and piano, in every crevice, hole, and corner, for at least an hour did I hunt without success. Eventually the hiding-place of the fugitive was discovered in the following singular manner: As I sat at my desk, I was startled by a mysterious noise which apparently proceeded from the interior of my 'Broadwood,' which, by-the-by, I verily believe knows something about the early editions of 'The battle of Prague.' The strings of this venerable instrument descend into ill-disguised cupboards, so that at the lower part there are two doors, or, in scientific language, 'valves.' On opening one of these, what should I see but the poor crab, who, at my approach, 'did' a kind of scamper polka over the strings. This performance I took the liberty of cutting short with all possible speed. On dragging away the performer, I found that his appearance was by no means improved since I saw him last. Instead of being ornamented with gracefully-bending polypes, he was coated, body and legs, with dust and cobwebs. I determined to try the effect of a bath, and presently had the satisfaction of seeing him regain his usual comely appearance. The next step was to replace him in his old abode; and having done so, I felt anxious to know how the creature had managed to scale his prison walls. The *modus operandi* was speedily made apparent; yet I feel certain that, unless one had watched as I did, the struggles of this little fellow, the determination and perseverance he exhibited would be incredible.

After examining his movements for an hour, I found, by dint of standing on the points of his toes, poised on a segment of weed, that he managed to touch the brim of the glass. Having got thus far, he next gradually drew himself up, and sat upon the edge of the vessel. In this position he would rest as seemingly content as a bird on a bush, or a schoolboy on a gate.

My curiosity satisfied, the *C. mœnas* was again placed in the vase, and every means of escape removed.

Here let me mention that I still had a Fiddler-Crab in my large tank, who had formerly lived in companionship with the shore-crab above mentioned. With 'the fiddler' I had no fault to find; he was always modest and gentle, and gave no offence whatever to my Blennies. He never attempted to embrace them, nor to usurp their lawful place at the table, nor even to appropriate their meals. On the contrary, he always crept under a stone, and closely watched the process of eating until the coast was clear, when he would scuttle out, and feed, Lazarus-like, upon any crumbs that might be scattered around.

Although so modest and retiring, I soon discovered

that this little crab possessed an ambitious and roving disposition. This made him wish to step into the world without, and proceed on a voyage of discovery—to start, indeed, on his own account, and be independent of my hospitality, or the dubious bounty of his finny companions. Taking advantage on one occasion of a piece of sandstone that rested on the side of the aquarium, he climbed up its slanting side, from thence he stepped on to the top of the vessel, and so dropped down outside upon the room floor. For nearly two days I missed his familiar face, but had no conception that he had escaped, or that he wished to escape from his crystal abode. It was by mere accident that I discovered the fact.

Entering my study, after a walk on a wet day, umbrella in hand, I thoughtlessly placed this useful article against a chair. A little pool of water immediately formed upon the carpet, which I had no sooner noticed, than I got up to remove the *parapluie* to its proper place in the stand, but started back in surprise, for in the little pool stood the fugitive fiddler moistening his branchiæ.

Taking up the little prodigal who had left my protection so lately, I soon deposited him in a vase of clear salt water. After a while, thinking it might conduce to the happiness of both parties, I placed him in companionship with his old friend, *Carcinus mænas*. This, like many other philanthropic projects, proved a complete failure. Both creatures, once so harmless towards each other, seemed suddenly inspired by the demon of mischief. Combats, more or less severe, constantly occurring, in a few days I separated them.

The 'fiddler' I placed in the large tank, where he rested content, and never again offered to escape— evidently the better of his experience. Not so his old friend, who still continued obstinate and miserable as ever. In his case I determined to see if a certain amount of sternness would not curb his haughty spirit. For two days I offered him no food, but punished him with repeated strokes on his back, morning and evening. This treatment was evidently unpleasant, for he scampered about with astonishing rapidity, and ever endeavoured to shelter himself under the granite centre-piece. When I thought he had been sufficiently chastised, I next endeavoured to coax him into contentment and better conduct. My good efforts were, however, unavailing. Every morning I placed before him a newly-opened mussel, but on no occasion did he touch a morsel. All day he continued struggling, as heretofore, to climb up the side of his chamber, trying by every means in his power to escape. This untameable disposition manifested itself for about a week, but at the end of that time, on looking into the vase, I saw the crab seated on the top of the stone, his body resting against the glass. I then took up a piece of meat and placed it before him. To my surprise he did not run away as usual. Having waited for some minutes, and looking upon his obstinacy as unpardonable, I tapped him with a little stick—still he never moved. A sudden thought flashed across my mind; I took him up in my hand, examined him, and quickly found that he was stiff and dead!

There is a little crab, *Porcellana longicornis,* or Minute Porcelain-Crab, frequently to be met with in certain localities.

The peculiarity of this creature is the thickness and the great disproportionate length of his arms, as compared with the size of his pea-like body. He possesses a singular habit which I have not observed in any other crustaceans. He does not sit under a stone, for instance, but always lies beneath such object with his back upon the ground; so that when a boulder is turned over, these crabs are always found sitting upon it, whereas the shore-crabs, when the light of day is suddenly let in upon them, scamper off with all possible speed; or if any remain, it appears as if they had been pressed to death almost, by the weight of the stone upon their backs.

The colour of *P. longicornis* is that of prepared chocolate, shaded off to a warm red.

Another crab, equally common with those already mentioned, is to be met with when dredging, and in most rock pools. At Wardie, near Edinburgh, I have seen hundreds of all sizes hiding beneath the rocks at low tide. Its scientific name is *Hyas araneus*, but it is better known as one of the Spider-Crabs. It claims close relationship with that noted crustaceous sanitary reformer, *Maia squinado*. Although this H. araneus is a somewhat pleasant fellow when you get thoroughly acquainted with his eccentricities, appearances are sadly against him at starting. Speaking with due caution and in the gentlest manner possible, consistent with truth, I must say that this crab is, without exception, one of the dirtiest-looking animals I have ever met with in my zoological researches. At a by no means hasty glance, he appears to be miraculously built up of mud, hair, and grit on every part, except his claws, which are long and sharp as those of any bird of prey.

The first specimen I ever saw, seemed as if he had been dipped in a gum pot, and then soused over head and ears in short-cut hair and filth.

The second specimen, although equally grimy, had some redeeming points in his personal appearance, for at intervals every part of his back and claws were covered with small frondlets of ulva, dulse, *D. sanguinea*, and other beautiful weeds, all of which were in a healthy condition. After keeping him in a vase for a week, he managed, much against my wish, to strip himself of the greater part of these novel excrescences.

Instead of minute algæ, we read that these crabs

are sometimes found with oysters *(Ostrea edulis)* attached to their backs. Mr. W. Thompson mentions two instances where this occurs, with specimens of *H. araneus*, to be seen in Mr. Wyndman's cabinet. Speaking of these, he adds, 'The oyster on the large crab is three inches in length, and five or six years' old, and is covered with many large Balani. The shell, a carapace of the crab, is but two inches and a quarter in length, and hence it must, Atlas-like, have born a world of weight upon its shoulders. The presence of the oyster affords interesting evidence that the Hyas lived several years after attaining its full growth.

For days after I had brought him home, my second specimen appeared as if he were dead, and it was only by examining his mouth through a hand lens that I could satisfy myself as to his being alive. When I pushed him about with an ivory stick he never resisted, but always remained still upon the spot where I had urged him.

This species of *acting* he has given up for some time, and at the present moment I rank H. araneus among my list of marine pets, for he does not appear any longer to pine for mud with which to decorate his person, but is quite content to 'purge and live cleanly' all the rest of his days.

The ancients imagined that *Maia squinado* possessed a great degree of wisdom, and further believed him to be sensible to the divine charms of music.

It is very curious, as well as true, that this animal has in a far higher degree than other crustacean, a gravity of demeanour, and a profound style of doing everything, that always excites our irreverent laughter, but at same time leaves an impression that, if justice were done, the animal ought to hold a higher position in the marine world than a scavenger and devourer of ocean garbage. If *Maia* and *C. mænas* be both eating out of the same dish, in the shape of an open mussel, the former seems ever inclined to admonish his companion upon greediness and want of manners. The only seeming reason why *M. squinado* does not really give such advice, is because of the impossibility of any individual speaking with his mouth full. The knowledge, too, that if he commenced a pantomimic discourse, it would give his young friend an opportunity of gaining too large a share of the banquet, may, perhaps, have something to do with his preferring to remain quiet.

As for *Maia's* possession of appreciative musical qualities, I can only state that both he and his friend *Hyas* really do convey to the beholder an impression confirmatory of this statement. I have frequently been amused to observe the singular phenomenon of each animal coming to the side of the vase and rocking his body to and fro, in apparent delight at the exercise of my vocal abilities, just as when a pleasing melody is being played in the concert room, we bend backwards and forwards, and beat time to the tune.

These animals also adopt the same course: it must be to unheard music (which the poets say is sweetest), that seems ever and anon to fall on their ears, giving them great delight.

The movements here alluded to may be in no way influenced by music; but such as they are, it is curious that they have not been noticed as an apparent explanation of the origin of the ancient belief regarding the Spider-Crabs.

A friend, on one occasion having procured for me, among other objects, a Common Limpet, I placed this mollusc in my aquarium, and soon had the pleasure of watching it affix its broad foot to the surface of the glass. After a while, on the Limpet slightly raising its canopy, I was surprised to observe a little Shore-Crab peer out from between the foot and shell. On suddenly ejecting the intruder by means of a small brush, he speedily hid himself from view among the surrounding pebbles. A few hours after, on again approaching the tank to view the Patella (which was easily identified, from the fact of an immense colony of Mussels being settled on its back), I found to my great astonishment that the crab had re-seated himself in his old position. I often repeated the sweeping operation, but without success, for the little rascal had become artful, and was not inclined to be driven forth a second time by a *coup de main*. I touched the Limpet frequently and saw it glue itself, as usual,. to the glass; but, singular to state, the creature always left a larger space between its foot and the circumference of the shell on the side at which the crab was seated, than on the opposite one, seemingly from a wish to accommodate its crustaceous friend. This space, moreover, let me observe, was larger than was absolutely necessary, for, as the shell was not air-tight, I was enabled to thrust my camel-hair pencil teazingly upon the crab, and was much amused to watch him clutch at the intruding object, and, at times, move about with it in his grasp, thus proving that he was by no means uncomfortably 'cabin'd, cribb'd, confin'd.'

For a whole week the crab remained in his favorite lodgings, and only resigned occupancy thereof when his friend gave up the shell—and died.

There is a certain species of crab, *Pinnotheres pisum*, or common Pea-Crab, frequently found in *Mytilus edulis*, the Oyster, and the Common Cockle. Indeed, one gentleman states, that on his examining, on two occasions, a large number of specimens of the *Cardium edule*, he found that nine out of every ten cockles contained a crab. Still, in no other instance than the one my own experience furnishes, have I ever heard of the Shore-Crab, or, indeed, of any other crustacean, becoming the guest of *Patella*.

The classical reader will not fail to remember Pliny's statement (somewhat analogous to that above narrated) of a small crab, *Pinnotheres veterum*, which is always found to inhabit the Pinna,—a large species of mussel. This latter animal being blind, but muscularly strong, and its juvenile companion quick-sighted, but weak of limb, the crab, it is said, always keeps a sharp look-out, and when any danger approaches, he gladly creeps into the gaping shell for protection. Some writers assert, that when the bivalve has occasion to eat, he sends forth his faithful henchman to procure food. If any foe approaches, *Pinnotheres* flies for protection with his utmost speed to the anxious bosom of his friend, who, being thus warned of danger, closes his valves, and escapes the threatened attack. When, on the contrary, the crab loads himself with booty, he makes a gentle noise at the opening of the shell, which is closed during his absence, and on admission, this curious pair fraternize, and feast on the fruits of the little one's foray.

For those of my readers who may prefer verse to prose, I here append a poetical version of this fable—equally pretty, but, let me add in a whisper, equally opposed to fact, at least in its principal details :—

> ' In clouded depths below, the Pinna hides,
> And through the silent paths obscurely glides ;
> A stupid wretch, and void of thoughtful care,
> He forms no bait, nor lays no tempting snare ;
> But the dull sluggard boasts a *crab* his friend,
> Whose busy eyes the coming prey attend.
> One room contains them, and the partners dwell
> Beneath the convex of one sloping shell :
> Deep in the watery vast the comrades rove,
> And mutual interest binds their constant love ;
> That wiser friend the lucky juncture tells,
> When in the circuit of his gaping shells

> Fish wandering enters; then the bearded guide
> Warns the dull mate, and pricks his tender side.
> He knows the hint, nor at the treatment grieves,
> But hugs the advantage, and the pain forgives:
> His closing shell the Pinna sudden joins,
> And 'twixt the pressing sides his prey confines.
> Thus fed by mutual aid, the friendly pair
> Divide their gains, and all their plunder share.'

There is one singular feature in the Crustacea which it may prove interesting to dwell a little upon. I allude to their power of living apparently without food, or at least without any other sustenance than is afforded by the animalculæ contained in the water in which they dwell. One accurate observer states that he kept a Cray-fish for a period of two years, during which time the only food the animal received was a few worms,—not more than fifty altogether. This statement I have often had ample means of verifying. Yet, on the other hand, strange to say, the crab is always on the hunt after tit-bits; and nothing seems to give him greater delight than a good morning meal, in the shape of a newly opened Mussel, Cockle, and above all—a Pholas. Let a youthful crustacean cast its shell, and rest assured, unless its companions have had their appetites appeased, they will endeavour to fall upon and devour the defenceless animal. This, to my chagrin and annoyance, I have known to occur repeatedly. When nothing else can be procured, not only the Lobster Crabs, but any Brachyurous Decapods who may be at hand, will set to work, and industriously pick off and eat the Acorn-

Barnacles attached to any object within reach. These facts show that the asceticism of the crab is not voluntary, and that when opportunity occurs, he is as fond of a good dinner as are animals possessed of a higher degree of organization.

It will be gratifying if other observers are able to verify the circumstance which I shall allude to hereafter, and which would seem to show that the *exuviation* of crustacea is expedited by affording specimens an unlimited supply of food.

'The organs for pursuing, seizing, tearing, and comminuting the food of the Brachyurous Decapods,' says Professor Bell, 'are carried to a high degree of development; . . . . these appendages consist of six pairs, of which some are actual organs of mastication, as the mandibles or the true jaws, the foot jaws or pedipalps, generally serving to keep the food in contact with the former, whilst it is being broken up by them.

'The buccal orifice in the Brachyura occupies the interior face of the cephalic division of the body, and is bounded anteriorly by a crustaceous lamina of determinate form, which has been termed the upper lip, and posteriorly by another, termed the lower lip. The mandibles occupy the sides of the opening. After these, and external to them, are the first, and then the second pair of true jaws, followed by the three pairs of pedipalps or foot jaws, the last of which, when at rest, close the mouth, and include the whole of the preceding ones. In the Macroura the pedipalps are very different in their forms, and have the aspect of very simple feet.

'The means of comminuting the food are not restricted to the complicated machinery above referred to, for the stomach itself contains a very remarkable apparatus, consisting of several hard calcareous pieces, which may be termed *gastric teeth*. They are attached to horny or calcareous levers, fixed in the parietes of the stomach; they are moved by a complicated system of muscles, and are admirably adapted to complete the thorough breaking-down of the aliment, which had already been to a considerable extent affected by the buccal appendages. These gastric teeth may be readily seen and examined in the larger species of Decapoda, as in the large eatable crab and the lobster; and it will be readily perceived how perfectly the different pieces are made to act upon each other, and to grind the food interposed between them.'

Having been on a certain day at the sea-side collecting, I was amused to observe the movements of two ragged little urchins, who approached near to where I stood, bottle in hand, examining some beautiful zoophytes by aid of a pocket lens. One of them had a short iron rod, with which he very dexterously hooked out any unfortunate crab who happened to have taken up its quarters in some crevice or beneath a boulder. Having captured a specimen, it was handed over to his companion, who quickly tied it to a string which he held in his hand.

I had seen many a rope of onions, but this was the first time I had seen a rope of crabs. On inquiry, I learned that the boys had taken two dozen animals in about two hours. When any of the green-bellied crabs happened to be poked out, they were allowed to escape back again as quickly as they pleased.

With poor *Cancer pagurus* the case was different, —every specimen, as soon as caught, being strung up, and doomed to 'death in the pot.'

The above, I need scarcely state, is not the usual manner of fishing for crabs, the approved plan being to take them in what are termed crab-pots, 'a sort of wicker-trap made, by preference, of the twigs of the golden willow (*salex vitellina*), at least in many parts of the coast, on account, as they say, of its great durability and toughness. These pots are formed on the principle of a common wire mouse-trap, but with the entrance at the top; they are baited with pieces of fish, generally of some otherwise useless kind, and these are fixed into the pots by means of a skewer. The pots are sunk by stones attached to the bottom, and the situation where they are dropped is indicated, and the means of raising them provided, by a long line fixed to the creel, or pot, having a piece of cork attached to the free end of the line; these float the line, and at the same time serve to designate the owners of the different pots—one, perhaps, having three corks near together towards the extremity of the line, and two distant ones—another may have one cork fastened crosswise, another fastened together, and so on. It is, of course, for their mutual security that the fishermen abstain from poaching on their neighbour's property; and hence we find that stealing from each other's pots is a crime almost wholly unknown amongst them.

'The fishery for these crabs constitutes an important trade on many parts of the coast. The numbers which are annually taken are immense; and, as the occupation of procuring them is principally carried on by persons who are past the more laborious and dangerous pursuits of general fishing, it affords a means of subsistence to many a poor man who, from age or infirmity, would be unable without it to keep himself and his family from the workhouse.' *

---

* Bell's Brit. Crus.

CHAPTER V.

Hermit-Crabs.

> 'Finding on the shoar
> Som handsome shell, whose native lord of late
> Was dispossessed by the doom of Fate,
> Therein he enters, and he takes possession
> Of th' empty harbour, by the free concession
> Of Nature's law—who goods that owner want,
> Alwales allots to the first occupant.'
>
> Du Bartas.

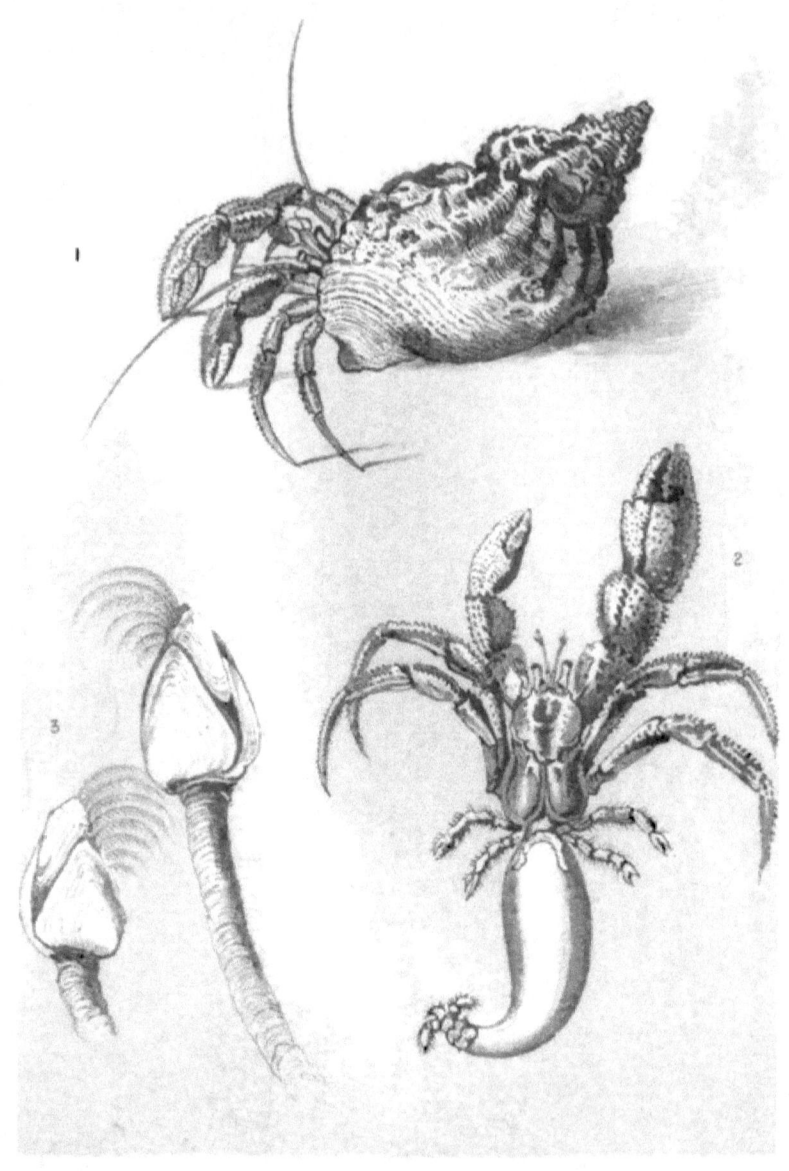

1 COMMON HERMIT-CRAB *(Pagurus bernhardus)* in shell of common Whelk.
2 Do Do out of shell.
3 SHIP-BARNACLES.

V.

TWICE in every twenty-four hours the waters of the ocean ebb and flow. Twice only in each month, however, do the spring-tides occur. For these there are few dangers that the ardent student of nature would not encounter. Lord Bacon tells of a certain bishop who used to bathe regularly twice every day, and on being asked why he bathed thus often, answered, 'Because I cannot conveniently bathe *three* times.' The zoologist, like the 'right reverend father' alluded to, would willingly undergo what appears to others much hardship and trouble, not only once or twice, but even three times daily, in pursuit of his favourite studies, did Nature but offer the kind convenience.

On these occasions the zoologist can pursue his researches at the shore, at a distance beyond the usual tidal line. Numerous boulders and rock-pools, during many days covered by the sea, being then laid bare and exposed to his eager, searching hands and eyes, he is frequently able to discover many rare objects, or at least, common ones revelling in almost giant-like proportions, and wonderful profusion.

The Soldier or Hermit-Crabs (to an account of whom we intend to devote this chapter), offer a most remarkable proof of this. Occupying the centre of a rocky excavation, I have repeatedly found several dozen of these comical creatures, each inhabiting the cast-off shell of a defunct Whelk (*Buccinum undatum*), which measured not less than five or six inches in length. To my surprise these aldermanic crustaceans possessed no companions of a smaller growth; while at a few yards nearer shore, as many shells would be found congregated together as in the more distant pool,—the largest, however, being no bigger than a damson, while the smallest might be compared to an infantile pea, or cherry-stone.

I cannot explain this appearance otherwise, than by supposing that the *Anomoura* become prouder, or, it may be, more cunning, as they grow older, and, having arrived at their full development, they fit themselves with their final suit; thereafter, in a spirit of aristocratic exclusiveness, they retire to fashionable subaqueous residences, distant as far as possible from the homes of the *canaille*, who inhabit the common, littoral boundaries of the shore.

The peculiarity, to which I alluded, of the *Anomoura* occupying shells that have formerly belonged to other animals, is so strange that some writers have not hesitated to express doubt upon the subject. This denial of a fact, which can so readily be proved, is one of the 'curiosities of literature.' Swammerdam, a Dutch naturalist contemptuously observes, 'What an idle fable that is which is established even among those who study shell-fishes, when they show some kind of the crab kind in their museums, adding at the same time, that they pass from one shell to another, devour the animals that lived in those shells, and keep them for their own habitations. They dignify them with the high-sounding names, and additions, as Soldiers, Hermits, and the like; and thus, having no experience, they commit gross errors, and deceive themselves, as well as others, with their idle imaginations.'

That there is nothing mythical in the matter can easily be made apparent to any person who chooses to visit the sea-shore. At such locality he need have no difficulty in recognising the Hermit-Crab, or meeting with numerous specimens for examination. Supposing such a one is at a rock-pool, and, moreover, that he knows by sight the Buckie (periwinkle), and Common Whelk, he will probably in such case be aware that the animals occupying these shells are snail-like in construction, and that their locomotion is consequently slow and formal. If, therefore, when peering into any pool he sees the Buckie, for instance, apparently change its nature, and instead of

'Dragging its slow length along,'

scamper off suddenly, or roll over and over from the

top of an eminence to the bottom, he may rest assured that the original inhabitant has departed, and that its place is occupied by a Lobster-Crab.

The cause of his strange peculiarity I will briefly explain.

In the true Lobster the tail forms a most valuable appendage. In the tail the principal muscular power of the animal is seated; and by means of it, too, the animal is enabled to spring to a considerable distance, and also to swim through the water at will. This important organ is well protected by a casing consisting of a 'series of calcareous rings, forming a hard and insensible chain armour.'

In the Lobster-Crab there is no such arrangement. 'The abdominal segment of this singular animal, instead of possessing the same crustaceous covering as the rest of his body and claws, is quite soft, *and merely enveloped in a thin skin.* To protect this delicate member from the attacks of his voracious companions, the poor Pagurus is compelled to hunt about for some Univalve, such as a Whelk or Trochus, and having found this, he drops his tail within the aperture and hooks it firmly to the columella of the shell. Why Providence has doomed the poor Hermits to descend to such physical hypocrisy, and clothe themselves in the left-off garments of other animals, it is not easy to conjecture. No doubt, besides the defence of their otherwise unprotected bodies, he has some other object of importance in view. Perhaps they may accelerate the decomposition of the shells they inhabit, and cause them sooner to give way to the action of the atmosphere; and as all exuviæ may be termed nuisances and deformities, giving to these deserted mansions an appearance of renewed life and locomotion, removes them in some sort from the catalogue of blemishes.'

Professor Jones, when treating of this class of animals, forcibly remarks that 'the wonderful adaptation of all the limbs to a residence in such a dwelling, cannot fail to strike the most curious observer. The *Chelœ*, or large claws, differ remarkably in size, so that when the animal retires into its concealment, the smaller one may be entirely withdrawn, while the larger closes and guards the orifice. The two succeeding pairs of legs, unlike those of the Lobster, are of great size and strength, and instead of being terminated by pincers, end in strong-pointed levers, whereby the animal can not only crawl, but drag after it, its heavy habitation. Behind these locomotive legs are two feeble pairs, barely strong enough to enable the Soldier-Crab to shift his position in the shell he has chosen ; and the false feet attached to the abdomen are even still more rudimentary in their development. But the most singularly altered portion of the skeleton is the fin of the tail, which here becomes transformed into a kind of holding apparatus by which the creature retains a firm grasp of the bottom of his residence.'

So great is the power of the animals to retain hold of their shell, and so intense their dislike to be forcibly ejected therefrom, that they will often allow their bodies to be pulled asunder, and sacrifice their life rather than submit to such indignity. This fact I have proved on sundry occasions. But supposing a crab to have taken a fancy to a shell, occupied by some brother Pagurus, (a circumstance of frequent occurrence), he quickly proceeds to dislodge the latter. Curious to state, this process never seems attended with any fatal result.

When watching the operation, it has appeared to me as if the crab attacked preferred to yield rather than be subjected to continuous annoyance, and the discomfort of keeping for so long a time buried within the inner recesses of his dwelling.

The contrast in appearance of the Hermit-Crab when seated in his shell, and crawling about minus such appendage, is great indeed.

This the reader will readily perceive by examining the Illustrations on Plate 4, which are drawn from nature, and are truthful portraitures of this singular creature.

I have already mentioned the extreme difficulty there is in expelling a Lobster-Crab. This, be it understood, applies only to the animal in good health; for no sooner does he feel sick than he instantly leaves his shell, and crawls about in a most pitiable plight. He sometimes becomes convalescent again by being placed solus in some fresh water, or laid out in the air for a few moments. But he ought, on no account, when in a sickly condition, to be allowed to hide himself beneath any pieces of rock or shadow of the Algæ.

If he is out of sight, be sure not to let him be out of mind; for, should he die in the tank, and his body be allowed to remain for any length of time, he will very soon afford you full proof that such toleration on your part is anything but pleasant.

Although, as already stated, this animal cannot be drawn out of his shell except by extreme force, the object can easily be obtained by aid of strategy. Having been for some time at a loss how to give certain young visitors a sight of the Hermit-Crab in his defenceless state, I, by accident, hit upon the following simple plan:—

With a piece of bent whalebone I lifted up a Pagurus, shell and all, and allowed the latter to drop upon the outer row of the tentacula of an Actinia, which quickly stuck fast to the intruding object. The crab at first did not seem fully alive to his critical position. He popped out of his shell and looked unsuspectingly around, until catching sight of my face, he instantly retired from view with a casket-like snap. In a minute he was out again, and this time prepared to change his position. For this purpose he gave several successive pulls, but finding all his efforts to remove his carriage unavailing, he unhooked his tail and scrambled down among the pebbles. My purpose was thereby gained, for the next moment he was resting in the palm of one of my juvenile friends, who seemed quite delighted with his prize. Twice afterwards, being in a mischievous mood, I gave the crab a fright in the way just mentioned; but it was quite evident, that what might be sport to me was death to him, for he was both annoyed and alarmed at my procedure. Even when guiltless of any intention of touching the creature, if I merely showed him the cane he immediately hobbled away at the utmost rate of speed he could muster. On several occasions I followed after and brought him back to the edge of the tank, although such conduct met with his strongest disapproval, and caused him for some time to sulk beneath an arch-way of rock work, away from the reach of vulgar eyes.

Upon the side and near the base of my tank a fine specimen of the Limpet was at one time attached. From the centre of its shell a forest of sea-grass waved gracefully, shadowing a large colony of Barnacles thickly clustered beneath. Soon the Patella decided upon taking its usual morning stroll in search of food, a task of little difficulty, standing as the animal already did upon the margin of a broad meadow, richly coated with a verdant growth, composed of the infant spores of the Ulva. Slowly moving along, the Patella, with its riband-like band of teeth, swept off the luscious weed in a series of graceful curves, thus making an abundant and healthful meal. Before proceeding far, however, he was forced to bear the weight of a Soldier-Crab, who had most unceremoniously climbed upon his back, and taken up a position at the base of the *latissima* fronds.

There seemed so much nonchalance about the Pagurus that I determined to watch his movements, and, if possible, to see how he would manage to descend from a position which, if the mollusc continued his mowing operations, would soon be unenviably high.

In about an hour the Limpet had reached the level of the water in the aquarium, and there took up his abode for the night. Next day and the next there was no change of situation. The crab now began evidently to perceive the danger of the position in which he was placed, for he constantly moved to and fro, and peered over into what must have seemed to him an unfathomable abyss.

While I stood, the Patella made a sudden movement of its shell—so sudden, indeed, as to startle its companion, who quickly put out his claws to save himself from falling. Unfortunately, in his spasmodic gesture he allowed the tip of one of his claws to intrude under the edge of the conical canopy, thus, in fact, pricking the fleshy 'mantle' of the animal within, who instantly, of course, glued itself to the glass with immoveable firmness. I suppose the same thing must have frequently occurred without my knowledge, for after a lapse of several days the Pagurus and his bearer were still in the same spot. I felt a growing alarm for the continued health of the Hermit-Crab, from the fact of its being poised so directly over the ever-expanded tentacles of a large Anemone. To prevent any mishap, I went to lift his crabship, with a view of transferring him to a place of safety, when, no sooner did he perceive the advancing forceps, than he rushed into his shell with a sudden and audible 'click,' forgetting for the moment that he stood on such ticklish ground. The consequence was that, seeking to avoid Scylla, he fell into Charybdis. In other words, he dropped plump upon the well-gummed tenter-hooks of the Crassicornis, which instantly closed and engulphed its prize. In vain did I endeavour with all speed to pick out the devoted Pagurus. The more I tried, the more firmly did the Actinia hold him in its convulsive grasp.

With extremely few exceptions, the Hermit-Crabs are always found to be a prying, prowling, curious class of animals, and are ever, like the husband of the fair Lady Jane—

'Poking their nose (?) into this thing and that.'

They will turn over each shell and pebble that comes in their way, and examine it with profound attention, or industriously climb up and roll down hillocks and trees in the shape of small rocks and sea-weeds, much to their danger.

I once possessed a Hermit-Crab, whose voracious movements afforded considerable amusement to myself and my friends. My Diogenes—or, as the Cockney news-boys used to pronounce the now extinct comic periodical, *Dodgenes*—on a certain occasion had climbed up a segmentally cut frond of Irish Moss. On reaching the topmost point, his weight became too great for the weed to bear; so, finding he was losing his equilibrium, in great alarm he made a clutch at the first object that stood near, in order to save him from falling.

A mussel was moored hard by, to the side of the vase by means of its silken byssus threads, and upon this friendly bivalve the Pagurus leaped by aid of his long taper legs. Unluckily the shell of the Mytilus was open, and the crab unwittingly thrusting his toe within the aperture, the intruding object was of course instantly gripped by the mollusc. This accident put him in a terrible fright. His gestures were most excited, and no wonder. Let the reader fancy himself hanging on to a window sill, at a height say of twenty feet from the ground, with the sash-frame fixed on his hand, and a huge iron foot-bath, or some such object, attached to the lower part of his body, and he will have a tolerably correct idea of the painful position of our crustacean friend.

After curling and uncurling his tail, and trying

several times in vain to throw his tub upon the valve of the mussel, he released hold of his encumbrance, and allowed it to drop. Although still hanging, he had no difficulty in rolling up his 'continuation,' and elevating his body to the walls of his prison. Once again upon solid ground, he laboured hard to get his leg free. But unsuccessful in his efforts, he adopted another course, and snapped it off in a rage.

Scarcely, however, was the act of mutilation finished, when the stupid animal apparently seemed anxious to recover his lost toe, (which I may mention, had in reality fallen down among the pebbles).

After scraping, then resting, and scraping again, many successive times, he at last succeeded in diving the points of his largest claw into the chasm formed by the gaping mollusc. Of course, the member was held as if by a powerful vice. Very soon his courage deserted him, and he seemed to wait and weep despairingly for fate to release him from the sad predicament into which he had foolishly fallen. Alas! he little knew the singular part that fickle fortune had doomed him to play,—to become, if I may so term it, a kind of Prometheus in the tank.

My pack of fishes, having been on short rations for several days, had become exceedingly ravenous, and consequently were keeping a sharp look-out for scraps. Hence their intense delight on catching sight of the devoted 'Dodgenes' can readily be imagined. Such a delicious *morceau* was perfectly irresistible:—

> 'Mercy, mercy!
> No pity, no release, no respite, oh!'

At it they went, 'tooth and nail.' First one and then another tore away a mouthful, until in the twinkling of an eye, almost, the martyr crab was left forlorn and dead—

> 'A remnant of his former self.'

During the early portion of last year I had a Hermit-Crab inhabiting a pretty Purpura, whose shell I wished to sketch as an illustration, it being of peculiar form and colour. On going to the tank I discovered that Pagurus had most apropos vacated his turbinated cot, apparently in consequence of his feeling rather squeamish. Thinking he might perhaps presently recover, or pick up another dwelling, I hesitated not to abstract the shell, in order to make the required drawing. I had not been occupied with my task for more than five minutes, when my attention was attracted by a great excitement and clatter pervading the tank. A hasty glance within the vessel sufficed to explain the cause of the hubbub.

The brief domestic drama of which I was a spectator, with its somewhat singular denouement, I will now proceed to unfold for the reader's entertainment. It conveys a good lesson in natural history, and also exhibits a striking example of life beneath the waters.

The Blennies, I may state, had become very voracious, pugnacious, and audacious; nothing seemed safe from their attacks. I had begun to feed them on the *Cardium edule* and Mussel, but such diet, after a time, only served to whet their appetite, which certainly appeared to 'grow by what it fed on,' for they darted about through the water in all directions, searching, as I suppose, for other dainties. These efforts were unsuccessful, until they caught sight of the plump, undefended portion of the body of their companion, the Hermit-Crab, who had just left his shell, as above stated.

The sight of such a feast must have (figuratively speaking) made their 'mouths water.' One after another these rascally fish dodged round the crustaceous victim, and gripped, and shook his 'continuation' with extraordinary violence. In vain did the crab try to act on the defensive; all his efforts to retaliate were ineffectual, and in this instance it might be truly said that 'might' overcame 'right.' He ran to and fro in great distress, scraping the pebbles and shells about (thus partly creating the clatter that I had heard while sketching), in the hope that he might find an empty univalve in which to deposit his mutilated carcase. When almost breathless and exhausted, he discovered a worn-out Wentletrap, and strove to lift his quivering body into the aperture, alas! without success. His strength failed him, and he fell dead at the very threshold of his new-found home.

While watching thus far the above transaction, the writer felt almost inclined to waver in the faith he had long held with others, namely, that fishes and other marine animals are insensible to pain. But the movements of this poor Hermit-Crab were as indicative of severe suffering as anything he ever witnessed in bird or quadruped.

Wishing to examine the remains of the crab, I stepped aside for a few moments to procure my forceps, but when I again reached the vase, to my intense surprise the defunct animal was nowhere to be seen! I could only account for so singular a circumstance by supposing some of the larger crustaceans had taken advantage of my absence to complete the work of destruction, and therefore took no further notice of the matter at the time.

I had often wished that some of my finny pets would deposit their spawn in the tank, and felt very anxious, if such an event did take place, that I might be near to witness it. But I was most anxious to watch the gradual development of the ova, and, if practicable, to become the fond owner of a host of infant 'fishlings.'

Guess the thrill of pride, then, which ran through my veins when, on peering into my mimic rock-pool, after a brief absence from home, I observed the largest of my Blennies to be apparently in an 'interesting condition.' I watched and petted her many times daily, and fed her with every suitable dainty that could be thought of. Sometimes I took her in the palm of my hand, and with a fine camel-hair pencil stroked her glossy back. This operation evidently gave great delight to the little beauty; and after a while, when my hand was laid in the water, she gently floated off into her native element with almost swan-like grace.

The law of nature being the same with this fish as with the Stickleback, I knew the nest, if there was to be one at all, should be built by the male. But as I could not detect any specimen of the 'sterner sex' among my pack, and there being no signs of preparation for the grand event about to take place, I felt in a manner compelled to carry out the nidifying task in my own humble way. Of course, I gave up all idea of 'weaving' a nest with bits of weed, stones, and marine glue; nor was such a style of structure a desideratum in the present instance, wanting, as I did, to take notes, in Paul Pry fashion, of the minutest particular that might occur within the building. The following was the plan I adopted. First was procured the exquisitely formed valve of a large *Pecten*, the interior of which was white and beautifully irridescent. This pretty cot, I said to myself, shall serve as a chamber for my *protégé*. The shell being deposited behind a piece of rock, in such a position that its side rested against the surface of the glass, I was thus enabled to watch what was going on within. Some fronds of sea-weed were trained around so as to form a kind of drapery. The Blenny, I am quite certain, knew perfectly well that all this care and preparation was on her account, for nearly the whole of each day she spent in the novel apartment extemporized for her accommodation. After a week had elapsed, she grew uneasy and pettish, was ever snapping at her companions, and hunting them about in all directions. On one occasion, however, she seemed to be uneasy, now dashing round the rock, then darting to the top of the tank, and down again upon the pebbles. Scores of times these movements were repeated, until I felt alarmed for her safety, and annoyed at my inability to relieve her sufferings. But aid from me being impossible, I felt compelled, though very unwillingly, to allow nature to take its course.

On looking into the aquarium one morning, I observed some strange object protruding from the fish. The little creature, too, on catching sight of me, came to the side of the tank, near to where I stood, and by her movements asked me, as plainly as any dumb animal could ask, to give her my assistance. After a few minutes spent in a 'brown study', I resolved to grant her petition, and immediately setting to work, drew from her—what? what do you suppose, reader? In truth neither more nor less than the body, head, and long antennæ of the 'martyr' Hermit-Crab! whose late sudden disappearance was now fully accounted for.

There are ten British species of Lobster-Crab, but one only, *P. Bernardhus*, to which the reader has been introduced, is common to our shores.

CHAPTER VI.

Exuviation of Crustacea.

(THE PHENOMENA OF CRABS, ETC., CASTING THEIR SHELLS.)

'As Samson at his marriage propounded a riddle to his companions to try their wits thereon, so God offereth such enigmas in Nature, partly that men may make use of their admiring as well as of their understanding; partly that philosophers may be taught their distance betwixt themselves, who are but the lovers, and God, who is the giver of wisdom.'—INTRODUCTION TO CONCHOLOGY, page 384.

VI.

The Armory of the Tower of London forms, it is generally admitted, one of the most interesting sights of the great metropolis. No one can look without wonder upon that goodly array of knights and noble warriors, nor help an involuntary sigh over the degeneracy of modern humanity. Though the figures before us are technically and irreverently termed 'dummies,' the hardened shell with which their body and limbs are cased we know has felt the throb of many a true English heart, maybe, glistened beneath the sun at Cressy and Agincourt, or perhaps on the bloody fields of Worcester and Marston Moor. It requires no great power of the imagination to transport ourselves to bygone centuries, and listen to the ring of hostile arms, the sepulchral voices of men whose heads are inurned in casques of steel, blended with the clash of battle-axes, the whizz of arrows, the neighing of steeds, the rattle of musketry, and at intervals the deep booming cannon's roar.

But, asks the gasping reader, what has this parade of mail-clad warriors and old battle-fields to do with so prosaic a theme as the exuviation of crabs? I must acknowledge that the question is a very natural one, for there appears at first sight no connection between the two subjects. The analogy will not, I believe, appear so forced when I mention my possession of a smaller, although hardly less singular armory, consisting of various coats of shelly mail, each of which, at one time or other, belonged to, and was worn by a living creature, and proved as effectual a protection in many fierce though bloodless combats as any casque or helmet worn by knight. Unlike the dummies of the Tower, my specimens are perfect, and give a complete representation, more truthful than any photograph, of the defunct originals, when armed by Nature *cap-a-pie*.

In plain words, I own a curious collection of the cast-off shells of various crabs, which have from time to time been under my protection. From the fact that no museum in the kingdom contains a single *series* of such objects, exhibiting the various stages of growth in any crustaceous animal, the reader will easily conceive the difficulty there must be in procuring them, and consequently the interest that attaches to the mysterious phenomenon of exuviation.

Strange to say, the subject of this chapter is one of the least known in the whole range of natural history. The facts connected with the process are few, and far from well authenticated. This state of things appears the more extraordinary, when we remember the great facility with which specimens of crustacea may be found.

For years past I have paid much attention to the elucidation of this subject, and during that period have had to submit to numberless mishaps and disappointments. For example, perhaps after watching a 'pet' day after day for months, anxiously expecting that exuviation would take place, in nine cases out of ten,—ay, in ninety-nine out of the hundred,—I would find that the process had been completed when I was asleep, or that the animal had died suddenly. In the latter case new specimens had to be procured, and the same watching process repeated, in most cases with the like unhappy results.

I will now, however, endeavour as briefly as possible to make the reader acquainted with what has already been written upon exuviation, as far as I have been able to learn, up to the present time, interspersing the narrative with such notes as may seem necessary by way of illustration, and then proceed, in the words of Shakspeare, to lay down my own 'penny of observation.'

The first clear and satisfactory remarks on this subject were made by the celebrated Reaumur, who lived above a century ago: 'The unexampled accuracy and truthfulness of this great naturalist is attested,' says one writer, 'by the fact, that of all the observations made by himself alone, far exceeding those of any other writer of past or present times, and occupying in their published form numerous large quarto volumes, scarcely one has been contravened by subsequent credible observers, whilst they have formed the substance of half the numerous compilations on insect life, acknowledged or otherwise, which have appeared since his time.'

Goldsmith, who derived his knowledge of this subject from Reamur, tells us, in his usual free and easy style, that crustaceous animals (as crabs and lobsters) 'regularly once a year, and about the beginning of May, cast their old shell, and nature supplies them with a new one. Some days before this necessary change takes place, the animal ceases to take its usual food. It then swells itself in an unusual manner, and by this the shell begins to divide at its junctures between the body and the tail. After this, by the same operation, it disengages itself of every part one after the other, *each part of the joints bursting longitudinally*, till the animal is at perfect liberty. *This operation, however, is so violent and painful that many die under it;* those which survive are feeble, and their naked muscles soft to the touch, being covered with a thin membrane; but in less than two days this membrane hardens in a surprising manner, and a new shell as impenetrable as the former supplies the place of that laid aside.'

This, then, was and is to a great extent, up to the present time, the universally adopted explanation. Goldie, of course, could not afford time, and it may be doubted if he possessed the requisite amount of patience, to confirm what he wrote by actual observation. Seeing that the statement was graphic in its details, and evidently either wholly or in part the result of personal observation, he very naturally gave it full credence. But what shall we say of a noted writer (Sir C. Bell)* who apparently half doubts the truth of exuviation, for although he mentions the particular account which Reamur gives, yet tells his readers that '*naturalists have not found these cast off shells.*' After such a remark as this, we need no longer sneer at the compilations of the author of the 'Vicar of Wakefield.'

I need hardly state, that at certain seasons of the year almost every rock-pool at the sea-shore will exhibit to the observant eye scores of 'these cast off shells' in a perfect state. The writer above quoted also remarks, 'We presume the reason that the shells of the crustacea are not found in our museums, is because they are not thrown off at once, but that the portions are detached in succession.' An ill-founded presumption this, the fact being that the inelastic integument is invariably (in all the Decapoda at least) thrown off entire, the eyes and long antennæ

Illustrations to Paley's Natural Theology.

sheaths, the claws with the hair attached, even the gastric teeth, all remain with wonderful exactness.

To look at the rejected shell, indeed, any person not previously acquainted with the fact would naturally suppose that he saw before him the living animal, a close inspection being necessary to dispel the illusion. As soon as the crab has emerged from its old covering, it increases with such astounding rapidity, that at the end of one or two days it can grow no larger until the next moulting time.

In referring to my own introduction to the subject of exuviation, I may be allowed to notice the annoyance a young aquarian experiences from the rapidity with which the tank water is apt to become opaque. As such a state involves considerable trouble, especially when the occupants of the tank are the subjects of continued observation, I may mention, in passing, that the means I adopted to correct this state of matters was either to syringe the water frequently, or what seemed to answer still better, to permit it to run off by a syphon into a basin on the floor.

When the opacity of the tank is occasioned by decaying animal matter, the only temedy is to remove the offending 'remains.' But with many of the common inhabitants of the tank—the crustaceans, for example—great difficulty is often experienced in ascertaining their state of health, with a view to sanitary investigation. As these creatures, instead of boldly exhibiting themselves during the day, generally hide under pebbles or pieces of rock, or are buried in the sand, it is sometimes necessary to submit the contents of the mimic rock-pool to a process of 'putting things to rights,' as the ladies say when about doing a kindness,—oh, horror!—to our books and papers.

It happened on a certain occasion that my aquarium was in an unsatisfactory condition. A nasty vapour arose from the base, and diffused itself over nearly the entire vessel. My fishes disliking their usual haunts, were all spread out at full length high and dry upon a ledge of rock work, projecting above the surface of the water. The little Soldier-Crab had managed to drag his body and heavy tail piece up the brae, hoping to breathe the fresh air in safety. His big brother was not so successful, and despite his efforts speedily came to grief. Finding he could not drag his carriage up the rock, he stepped out of the lumbering vehicle. His appearance soon became woe-begone in the extreme. In a few minutes he expired. The buckies, too, with singular instinct, had collected in a row along the dry ledge of the tank.

Upon counting the numbers of my little colony, I found all right excepting *C. Mænas;* him I could not discover, and I soon began to suspect that he was defunct. No time, therefore, was to be lost, so a diligent search for his remains was instantly commenced.

Fishes, Buckies, Hermits, &c., were speedily placed in safety in an extemporaneous tank—nothing else than an old pie-dish. This receptacle, when partly filled with sea-water, admirably answered the required purpose.

The water in the large vase was gently run off, and on approaching the base I found, as I expected, the dismembered carcase of the crab. One leg lay here, and another there, while the body was snugly esconced beneath a stone, on which sat my favourite limpet with its curiously formed shell, profusely decorated with a plume of sea-grass and infantile *D. sanguinea*. Here, then, I thought, was the mystery explained. It was from this spot that the noxious vapour must have emanated. Of course, the body of the crab was removed; but in performing this necessary act I tilted the stone, and so disturbed the Limpet. Guess my surprise at observing the overturned shell of the Patella to be quite empty, and its former occupant lying before me a mass of putrefaction.* It now began to dawn upon me that I must have libelled *C. Mœnas*. A few moments served to confirm this opinion, for on lifting the stone, there darted out *a*—I could scarcely believe it was *the* crab, who instantly went through a circus-like performance around the circumference of the vessel.

---

* This affords an important hint to the young aquarian to watch the Patella, and occasionally to touch its conical house, to make sure the proprietor is alive and well.

The reader will be prepared to learn that what I had at first observed were portions of the exuvium, which had by some means been distributed over the tank.

Many months did I wait with nervous anxiety to see the exact process of exuviation, but, except in the instances I am now about to chronicle, my wishes were never gratified.

I had at one time in my possession six little vases, each containing a crab measuring about one inch across the back (*carapace*). By constant watchfulness, morning and evening, for several months, I naturally entertained a confident hope of being favoured with a sight of the moulting operation in at least a single instance. But no; persevering though my endeavours were, I was always disappointed. The exuviæ were cast regularly enough, but the crabs so managed matters, that the process was completed either when I was asleep, or had just gone away. I could almost have sworn that the whole pack had entered into a league to annoy me.

On one occasion I sat up all night, feeling confident, from symptoms which a certain Cancer Mænas exhibited, that he was speedily about to exuviate. Alas! I was mistaken. On my endeavouring to expedite the event by lifting up the carapace of the crab, I received a nip on my finger so severe, that I shall never forget it.

But at length in the early portion of last year

(1859), I, most happily for my own peace of mind, did actually witness the entire process of exuviation in a tolerably large specimen of the Common Shore Crab. The animal in question, who was domiciled in a crystal vase, or, in common language, a glass tumbler, rendered himself a favourite from his constant habit of poking part of his head and his entire claw (he had got but one), out of the water whenever he caught sight of me. Who could resist such a powerful, though silent appeal to 'the generous impulses of one's nature' as this? Certainly I could not, and therefore, once a day at least, gave Master Cancer the half of a newly-opened mussel, a tit-bit that was greatly relished. He would sometimes get a grip of the valve, and allow himself and the Mytilus to be entirely raised out of the water. Improving upon this, he would then partly finish his meal while seated in my hand. On the morning of the above mentioned eventful day, I gave the crab a portion of a Pholas, but to my surprise, the heretofore high-class dainty remained untouched. I was in ecstasies! for I felt morally certain that the grand event, so long looked for, was soon to take place. Consequently, I took out the crab, cleaned the windows of his dwelling in order that I might the better see what was going on within, treated him to some fresh water, as well as a new frond of sea-weed, and then again introduced my pet to his old apartment.

THE PROCESS OF EXUVIATION.

Before doing this I had the animal closely examined, to see if any signs of the approaching moult could be detected, but none were visible, except that the glassy bags, if I may so call them, which for some weeks had been gradually thrown out from the stumps of the three mutilated limbs, appeared finer in texture than usual. Indeed, so transparent had they become, that I could distinctly see the contour of the new limb about to be reproduced, folded up within each capsule.

A few minutes after the crab had been placed in the tumbler, I gave a peep to see how he was getting on. To my intense surprise, I observed that his shell had just opened near the tail! My first feeling was one of sorrow, thinking that in handling the specimen I had been too rough, and had perhaps injured it. This apprehension was soon changed to delight, as I became by degrees aware that exuviation had actually commenced.

The operation did not extend beyond five minutes (although the time appeared much longer to me), and was carried on by gentle, and at first almost imperceptible degrees. The shell, or carapace, was slowly raised over the back, and gave one the idea of the rear view of a lawyer's white wig when tilted over his brow, thus exposing the natural black hair on the occiput below; for, as the body of the animal came forth, it was very dark in colour, while the old case assumed a whitish hue. I need hardly say, the leg sheaths of the crab did *not* split open, and yet the corresponding limbs were drawn out with the greatest ease. Moreover, they did not appear in view one by one, but in a cluster, as it were, and packed close to the bent body of the crab.

During the entire process the animal appeared to use scarcely any exertion whatever, certainly not half so much as any human being would exhibit in throwing off the most trifling garment. In fact, the crab seemed to swell painlessly, and gently roll or glide out in a kind of ball. Until it had completely escaped from its old shell, I was somewhat puzzled to guess what shape it would eventually assume. The eyes and antennæ, so soon as they left their old sheaths, commenced, together with the flabellæ, to work as usual, although as yet they were still inside the exuvium. This circumstance was distinctly visible by looking through the side of the half-cast shell.

It was a curious and extraordinary sight to see the eyes gradually lose their brilliancy, and exhibit the filmy, lack-lustre-like appearance of death, while the act of exuviation was being accomplished. I may add that the tumbler which held my little captive stood upon a table near a large window, and that the sloughing operation was watched through a powerful hand lens.

On an after and well-remembered occasion, I saw a moderate-sized Partane standing on the top of a bush of *Chondrus Crispus* that grew in my aquarium. The fronds were attached to a piece of sandstone, placed uppermost upon a cluster of rock-work, situated, as before mentioned, in the centre of the vessel, and rising slightly above the level of the water. Thinking he was planning means of escape, I turned away for a few moments to procure a simple instrument wherewith to carry him to a less elevated position. On my return I saw him in the act of backing out of his shell. It was a singular circumstance that I should have just risen from the perusal of a talented author, who informed me that 'the crab hitches one of its claws into some crack or fissure, and from this point of resistance gives more power in emerging and withdrawing itself from between the carapace and the tail.'

Certainly no statement could more inadequately describe what I had witnessed in both of my crabs. Not only was the whole operation performed with perfect ease, but I am much inclined to believe with a degree of pleasure. For a while one of my crabs stood in juxtaposition to the shadow of its former self, and rubbed his antennæ and wee peeping eyes as if awakening from a sleep. He had been lately, there was no doubt, living in an oppressed state, and might probably have surveyed things around him somewhat darkly, but now all was bright and clear again. On turning, the first object that caught his awakened eye was his cast-off vestment, which he seemed to scan as dubiously as a grown man would an exhumed pair of boyish corduroys, and mutter musingly, while stroking his chin, 'Well, come what will, it can never be my *case* again.'

On taking it in my hand, the Partane felt quite soft and velvetty to the touch, and exhibited no signs of alarm.

Since then I have repeatedly had shells of crabs cast *in smooth glass globes, containing nothing else but clear salt water.* This fact, in my opinion, completely subverts the statements of certain writers, who assert that these animals require extraneous assistance when about to exuviate.

Some writers have questioned the truth of the generally-received opinion that the new parts of the crab are derived from the old: that, for instance, a claw is regenerated within a claw, a limb within a limb, eyes within the eyes, and that on exuviation each is withdrawn from the pre-existing organ as from a sheath. But my operations tend fully to confirm the popular and existing belief.

There is yet one curious point connected with this subject which requires explanation, as it is not generally understood. I allude to the apparent disproportionate smallness of the 'glassy bag,' situated at the stump, as compared with the size of the regenerated limb, which is supposed to be folded up within the bag previous to exuviation. On looking at the newly-formed member, we can scarcely believe it possible that the transparent case could by any possibility have held it. The mystery vanishes if the new limb or claw be examined; for, although in shape it is perfect, even to the most minute particular, it remains for a certain **period** comparatively useless to the animal, **from the fact of** its **being** utterly devoid of flesh.

The new limb, therefore, can be considered **merely** as an **expanded case**, which, by a wonderful **law** of nature, **becomes** slowly filled up and completed. Immediately after exuviation has taken place, and a **claw is** introduced in the place of some mutilated stump, if any one will pull off the new member, he can readily confirm the truth of what I have stated, and, moreover, be able to test into **how very small** bulk the new limb may be rolled.

As **the** reader may remember, Goldsmith states that **the crab casts its** shell 'regularly once a-year, at the **beginning of May.'** Professor Owen fixes the date in **the month of August.** Professor Bell states, that 'there **is no** doubt exuviation takes place *annually* with great regularity, until the growth is completed, which, **in** many species, is not before the animal is many years old.' Another professor, treating on the same subject, thus writes, 'We are told that all this coat of mail is *annually* thrown off in a single piece by the contained animal,—the great proficient in Chinese puzzles may well be posed at this greater puzzle.' In fact, all writers whose works

I have had opportunity of examining repeat the statement. Mr. Ball, who writes from personal observation, apparently confirms beyond a doubt, the annual moult of Crustacea. This gentleman, we learn, kept a Cray-fish alive for two years in a vase, and found that *during each year its exuvium was shed but once.*

It may readily be believed, with such a formidable array of contrary evidence, that I offer my own observations with modesty. But at the same time, I feel justified in confidently stating that the moult of the crab, (in its comparatively youthful state, at all events), takes place not only once, but many times during each year of its existence. My specimens may, perhaps, be considered exceptions to the general rule, but the facts I relate cannot by any possibility admit of doubt. The cast-off shells lie before me as I write.

Here is a set of three belonging to the same animal, exhibiting with marvellous exactness the gradual development of a broken claw. In the first the member appears very diminutive, in the second it is nearly twice its size, while in the third it has advanced to its natural form and bulk. To my regret, I cannot state the exact period that elapsed between each successive moult, but I am confident that the trio were cast in the course of a very few months.

I may here take the liberty of informing the uninitiated, that the appearance of the above objects is extremely pleasing; for, as the exuvium becomes dry, its colour changes to a bright scarlet, somewhat resembling that which the crab assumes when placed for a time in boiling water.

The next series of specimens, five in number, possess even still greater interest than the first examples. They were produced by a youthful *C. Mænas*, at the following consecutive intervals:—

The first moult took place on 11th April 1858; the second on the 22d of May following; the third on July the 3d; the fourth on the 30th of August; and the fifth on the 26th of September in the same year. So that between the first and second period of exuviation there was an interval of forty-one days, between the second and third forty-two days elapsed, between the third and fourth fifty-eight days, but, singular to state, between the fourth and fifth moult *only twenty-seven days intervened.*

My first impression was, that as the creature grew older, its shell would be renewed less frequently, and the dates of the sloughings seemed to support this idea—until the fourth moult. It had occurred to me that perhaps the operation might be accelerated by the amount of diet which the crab consumed. In order to test this, I fed the animal carefully every day, as though he were a prize beast about to be exhibited at some Christmas show. Nothing loath, he ate of everything that was placed before him with a gusto that would have done credit to an alderman. The result was, that the shell was renewed in less than half the time that elapsed between the preceding moults.

These interesting investigations, which had been conducted thus far so satisfactorily, were suddenly brought to a close by the death of my protégé. This sad event occurred unexpectedly, not from over-feeding, as some persons may suppose, but from natural causes.

Whether increase of food always produces a like effect to that mentioned, is a point that I hope some of my brother naturalists will be able to determine. That the moult was accelerated by such means in my own specimen I have not the slightest doubt, for, on no other grounds can I explain its unusually speedy occurrence.

I may here assure my readers that the above dates may be confidently relied upon as correct, and also that each exuvium was produced by the same crab—one specimen only being in the tank during the whole period.

Since the foregoing was written, I have again been fortunate enough to have ocular demonstration of the phenomenon of exuviation, as occurring in a *Cancer Pagurus*, about as large as a moderate-sized walnut.

While watching this crab, it flashed across my mind that it would be a happy circumstance if by any means *I could arrest the process then going on before my eyes, while it was yet only half completed,* in order that others might also be enabled to witness the marvellous act of exuviation.

But how to carry out this scheme was the rub. I knew that—

> 'If 'twere done, then 'twere well it were done quickly.'

One minute passed,—two minutes flew by ;—the crab would speedily complete his labours ; still was I perplexed.

To plump it into fresh water would, I knew, be fatal to the animal, but not in such a speedy manner as was desirable. Boiling water next suggested itself, and doubtless would have answered the purpose effectually, had a supply been near at hand at the time, but such was not the case. I then thought of *spirits.* Ah ! capital idea. . . . . Before the third minute had passed, I might be seen to rush frantically to the sideboard, pour *something* into a glass, then dart back to the tank, dive down my trembling hand, bring up the poor unfortunate crab, and drop it into a fatal pool of pure "Glenlivet."

The animal appeared to die quickly, and was next day transferred to a vessel filled with Mythilated spirits. As it luckily turned out, the whisky answered the intended purpose remarkably well.

The preparation in question is, as far as I can discover, *unique;* at least I have neither read nor heard of another such existing in any private or public museum in the kingdom.[1]

It shows at a glance the increase that instantaneously takes place in the size of the crab after the act of exuviation is performed, the portion exuded being on a scale considerably larger than the old covering, which, however, is capacious enough to hold that half of the animal that had not effected its deliverance at the moment when the novel arrestment was so unceremoniously served.

The fourth and fifth pair of legs are free, while the eyes and antennæ are also drawn out of their sheaths. (This is not very evident now, but such is really the fact, I having distinctly seen those organs in motion when the animal was in the living state.) The *chelæ*, or large claws, being still undetached, serve to bind the crab to its old integument, and thus enable the act of exuviation, or one phase of it at least, to be distinctly apparent.

I know of no work on Natural History that speaks of the Hermit-Crabs (*Anomoura*) casting their shells, and on this account I have given some attention to them. These animals being so common, I kept by me at least a dozen specimens for the purpose of observing some of them, if possible, in the act of

---

[1] A drawing of this crab will be found on Plate 3.

exuviation. The result of my labours has not been so satisfactory as I could wish, from my not having been able to collect any 'sets' of exuviæ. I cannot, therefore, speak with certainty as to the frequency of this phenomena. By this time my readers will know that the tail of the Hermit-Crab is very tender and fleshy, being covered merely with an extremely delicate membraneous skin, while the carapace, claws, and antennæ of the animal are protected by a hard crust, similar to the Lobster, Cray-fish, &c.

From this peculiar formation of the crab, I was not at all surprised to find, on several occasions, the upper part of its body alone cast off, and therefore came to the very natural conclusion, that as the tail was soft, it would grow and increase in proportion to the other parts of the animal, without ever needing the skin to be changed.

Each morning and evening during the time my experiments were being conducted, I examined all the tanks attentively, to see whether an exuvium had been cast. If visible, the object was picked out and gummed in a box, and a date placed above it for future reference. After having performed an operation of this kind one afternoon in October 1858, I saw a Hermit-Crab (who had cast his shell on the previous day) hurriedly leave his testaceous dwelling, then scrape away at his tail, and after a moment's interval, leap into his old seat again. On inspection, I found to my surprise *that he had actually* *slipped off the skin of his tail!** much in the same fashion as we would draw off a well-fitting glove. Here was a strange and unexpected discovery.

On submitting the exuvium to the microscope, we find that the covering of the false feet, and the cilia attached to the same, all remained fixed in their natural position to the tail-piece. Although in several cases I have had no difficulty in discovering the rejected cuticle of the tail, at other times it has eluded my search. The cause of this I cannot explain. It may be that the animal, adopting the habits of the toad, swallows a portion of its exuviæ as soon as cast. On two occasions I found the slough of the body and claws of a crab, and waited patiently for several days, without success, expecting to get the tail portion. Growing impatient, it occurred to me that it would be a curious experiment to try and draw off the exuvium with my fingers. This was easy to talk about, but difficult to perform.

In the first place, the crab would not, if he could help it, allow himself to be handled even in the most gentle manner. To overcome this difficulty the shell had to be broken. This was done; but, alas! the shock nearly killed the poor little Hermit. After some trouble, I carefully unwound his body from the whirls of the Top Shell, and proceeded to perform the intended operation. Reader, have you

---

* The fact of the exuvium of the Hermit-Crab being cast off in two pieces, and at different periods, I have since confirmed 'many a time, and oft.'

ever seen a child take a rose-bud in his hands, and force open its half-pouting blossom, in the belief that by so doing he was assisting nature? If so, you must have watched the puzzled expression of the boy's countenance when he beheld the leaves fall one by one at his feet, and the bud itself exhibit evident signs of approaching decay.

In just such a position did I stand with regard to the poor Hermit-Crab, for, in spite of all my care in manipulation, the skin of the animal was so tender and delicate that the first gentle pinch caused a puncture which proved fatal; and as to drawing off the covering, the thing I now believe to be impossible, even under the most favourable circumstances.

The upper portion of the Soldier-Crab, I may mention, is cast off in one piece, while the animal is seated in its turbinated dwelling. The act is performed with the most perfect ease. Unlike the *Brachyura*, the *Anomoura* do not exhibit signs of such rapid growth immediately after exuviation, but increase in size very gradually indeed.

CHAPTER VII.

Prawns and Shrimps.

'Men holden ye therefore prophanes,
Ye eaten neither shrimps nor pranes.'

VII.

ALTHOUGH abundant at many parts of the Scottish coast, at Cockburnspath (situated near the mouth of the Frith of Forth) only, has the writer met with the very beautiful prawn, *Palæmon Squilla*. At this locality specimens were very frequent in rock-pools situated near the shore, nor were such difficult to capture. The small net being placed cautiously over their head, the animals did not show signs of resistance, until they found themselves, by a sudden jerk of the hand, drawn bodily out of the water. Then, indeed, unless some degree of skill was used, the captives would give a powerful spring, and escape, from the confinement of the net, to the more congenial element from which they had been so unceremoniously ejected.

Prawns (*Palæmonidæ*) exuviate very frequently, in some instances as often as once or twice a month. No sooner is one coat thrown off, and the Palæmon recovered from the weakness which the process occasions, than it commences, at first at intervals, and then almost incessantly day and night, preparations for a renewal of the wonderful operation.

Every part of the body—eyes, antennae, and especially the sub-abdominal fins, to certain portions of which the ova are attached in groups, and the lobes of the tail—are submitted to a severe rubbing and brushing process. The appearance of the prawn at this period is really most interesting, and, I may also add, amusing. Sometimes the tail is compressed inwards, beneath the body for a few seconds, and then suddenly elevated and forced out with donkey-like extravagance of gesture, the animal the while standing upon its first pair of forcep-like feet.

At the appointed time the shell opens at the back part of the head, and the prawn becomes gradually freed from its old covering. The marvellous process completed, like all its crustaceous brethren, the creature becomes to a certain extent helpless, and if such convenience be afforded, retires for protection beneath some shell or fragment of rock, from whence it soon re-appears, and repeats its gymnastic exercises, which cease, however, for a few days, as soon as the new coat is sufficiently hardened.

The prawn is an extremely interesting occupant of an aquarium, from the fact of its being constantly on the move, and also on account of the pretty blue and orange markings of its many-jointed legs, and the singular transparent appearance of its body.

This latter feature is made still more notable when the animal happens to have the ova attached, as the latter are opaque, and of a deep brown colour approaching to black. By the prawn the act of exuviation seems to be considered an event of no slight importance, and, although occurring so frequently, is fraught with danger. Specimens oftener die at the moulting time than at any other. In fact, unless I am very much mistaken, they are then subject to some peculiar disease, which is apt to prove fatal. At all events, several of my little captives, after having performed their gymnastic movements (before alluded to) for several days, turned sickly, and died. The commencement of their illness was always denoted by a small, white, opaque dot that mysteriously appeared in the centre of the body. This object speedily increased in size, until it eventually spread over the entire animal. Then, no longer diaphanous, the flesh of the prawn seemed composed of a solid substance not unlike lime or pounded chalk.

The Shrimp is so common, and so well known, that a lengthened description of it is unnecessary. I shall, therefore, merely record an ingenious plan by which specimens of the *Crangon vulgaris* may be procured by visitors at the sea-side, who do not care to wade in the water with a large net.

It is one generally pursued by Scotch boys as a mere amusement, for neither shrimps nor prawns are eaten to any great extent by the inhabitants of Scotland generally.

On arriving at a pool, a person will soon know whether shrimps are contained therein, from the number of sand clouds that are raised by these little crusty fellows at any intrusion upon their privacy. Many persons employ a hand net, and pass it rapidly through the water, thinking thereby to startle and entrap the animals in question. Sometimes the plan succeeds, but more often it turns out a failure.

Instead of using the net, let the young zoologist stoop down, place the palms of his hands suddenly upon the surface of the sand, then slowly draw them near each other, at same time cautiously close the fingers, and he will in all probability feel the objects of his search wriggling to escape from his unwelcome and unfriendly grasp.

To satisfy curiosity, take one of the captured specimens and drop it in the sand that surrounds the cavity in which your skill as a shrimper has been exercised, and I will venture to assert that, in an instant, the little creature will have disappeared as if by magic—such is the wonderful rapidity with which the shrimp burrows itself. Even when lying upon the surface a practised eye is required to detect the presence of a shrimp, in consequence of its colour being of the exact shade of the sand in which it hides. In clear pools its body is of a light drab colour, which becomes changed to a dark tint when the animal is located in a pool, the base of which is of a sombre hue.

The prawn, or shrimp, is somewhat of a gourmand, and requires to be fed occasionally. The most simple food to give either, when in an aquarium, is an open mussel or cockle. A marine worm, such for instance as the *Terrebella*, however (as on one occasion I vexatiously discovered), is a dainty more highly prized than the flesh of a bivalve, but one which cannot often be indulged in from its comparative rarity.

CHAPTER VIII.

Acorn Barnacles.—Ship Barnacles.

'Barnacles turn Solan Geese
In the islands of the Orcades.'

VIII.

If the reader has been struck at what has been said in regard to the exuviation of crabs, &c., he will probably be more surprised when I state that precisely the same phenomena take place in the simple *Acorn Barnacle*, that studs in countless numbers almost every rock and shell situated between tide marks. No one can visit the sea-shore, at certain localities, without noticing the white spots which constitute the shells of the cirripeds in question, although he may not be acquainted with the marvellous beauty of the animal contained within each.

Its loveliness, it is true, is in nowise apparent when parched and dry; but let the welcome waves advance and playfully dash their spray against the dwelling of the little crustacean, and quickly its valves will open, displaying a delicate feathery plume, thrust forth and hastily withdrawn again.

As it is not convenient to watch the movements of this animal in a rock-pool, let me request the reader kindly to take a peep into my aquarium.

Here is a Trochus shell, for example, inhabited, as you perceive, by a Soldier-Crab, the surface of which is thickly covered with shelly cones, of small dimensions. These are the Barnacles (*Balani*). The Trochus most fortunately being near the side of the glass, is capitally situated for our purpose. Take the hand lens, adjust its focus, and watch carefully for the opening of the cones. Tush! The hermit never *will* rest contented in any position for two consecutive minutes; but see! as he walks away the fairy hands are being rapidly thrown out and made to sweep the water in graceful curves, thereby suggesting some resemblance to a bevy of school children at Christmas time, bidding *adieux* to their friends, while seated on the roof of an old stage coach.

Carefully I lift the Pagurus bodily out of the tank, and transfer him to a wine glass filled with clean water. After a few minutes have elapsed, the hands again commence their fishing operations. Observe, now, that these organs fan the fluid in such a manner as to catch any animalculæ that may be near, and draw them towards the aperture caused by the opening of the valves of the Barnacle. A close inspection will, I am sure, prove to your satisfaction that there is also distinctly apparent a second and smaller cluster of feathery fingers, whose duty it is to catch the food, brought near by the larger and corresponding organs, and finally convey it into the mouth of the little cirriped. There may be, in the wide range of Nature's lower scale of life, prettier sights to gladden the eye of the student than that above described,—but if so, I must confess *my* inability to indicate where such are to be found. The fishing apparatus here mentioned consists of a number of slender *cirri*, thickly coated with microscopic filaments (cilia), and is, at certain periods, thrown off complete and entire by the process of exuviation, just as we have seen it occur in the higher crustacea.

Would you, my young friend, like to procure an exuvium of the Barnacle for examination? Yes. Then follow the directions I am now about to give, and your wish will be speedily gratified.

Presuming that your tank already contains a number of Barnacles attached to various objects, and that such have been in the same vessel for some weeks; syringe the water for a few minutes, and you will find floating about, or rising to the surface, many specimens of the desired object. They will, in all probability, be visible to the naked eye. To attempt to lift one out of the water, however, by means of your finger and thumb would be utterly useless. Such a procedure, even were it successful, would inevitably mar the delicate beauty of this ' inessential' object, which, spirit like, casts no shadow upon weed or water. The best plan is to insert a tube of glass into the aquarium, in such a way that the exuvium may ascend the interior. Then place your finger on the top, and draw the tube out of the water, and you will be able to deposit the skin of the Barnacle upon a slip of glass by merely lifting off your finger. The specimen can then be leisurely arranged, and spread out by aid of a hand lens and fine pointed needles.

Walking by the sea shore-one fine summer afternoon, I met a fisher boy running along with some curious objects spread out in the palm of his left hand, while in his right, suspended from finger and thumb, appeared a still more desirable prize.

At first glance I detected the objects to be specimens of the *Lepas anatifera*. They had, so the boy stated in answer to my inquiries, been plucked from the base of a ship newly arrived from a long voyage. When I offered him sixpence for the 'lot,' the embryo plougher of the deep looked up in my face with a singularly mistrustful expression, and said, 'D'ye mean it, sir?' I gave speedy assurance of my sincerity, and on receiving the purchase money, after handing over the Barnacles to my custody, the young urchin started off as fast as his legs, encased in huge wading boots, would allow him. His alarm was quite unnecessary, for although in a few days after I would not have given a penny for a thousand, I would willingly, on the above occasion, have paid five shillings for a single specimen, rather than have missed the opportunity of possessing such an interesting object as the Ship Barnacle.

On placing them in water one only of the creatures showed any signs of life, and by next morning they made the scentral organ of my face so highly indignant that, in order to allay its irritability, I was obliged to remove the defunct animals to the outside of the window. There they remained for several months, and were eventually transferred to the privacy of a card-board box. Although twelve months have elapsed since the last-mentioned removal took place, these creatures even now, when the lid of the case is lifted, give out a smell, so 'antient and fish-like,' that I believe not a few of 'the sweet perfumes of Arabia' would be needed in order to subdue its power.

One cluster contained thirty Lepades, and the other eighteen. The average length of each Barnacle is about three or four inches. One, however, measured nearly ten inches. The fleshy stalk is of a purplish-grey colour, semi-transparent, and perfectly smooth. The shell, which consists of five pieces, is bluish-white, while that portion from whence the cirri protrude appears of a brilliant orange, the cirri themselves being exquisitely tinted with violet, shaded off to a deep purple.

I may here mention that the above animal was by our ancestors most unaccountably supposed to be the young of the solan goose!—a bird that haunts in vast numbers the Bass Rock and Ailsa Craig. Indeed, a common belief in different parts of Scotland, and over the west of England was, that the shells grew upon certain trees, and in process of time opened of themselves; whereupon a certain animated substance contained within the shell dropped down, and according to the place where it fell perished or fructified. By falling into the water it grew to be a fowl; but by falling upon land the vital principle became extinct. The fowls which resulted from the more fortunate contingency were called Barnacle Geese in Scotland, and Brant, or Tree Geese in England. This delusion appears to have arisen from the fact of Barnacles having been found in great abundance on trunks and even branches of trees long submerged in the sea.* Bishop Hall thus alludes to the popular notion in his Satires:—

> 'His father dead! tush, no, it was not he;
> He finds records of his great pedigree;
> And tells how first his famous ancestor
> Did come in long since with the Conqueror.
> Nor hath some bribed herald first assigned
> His quartered arms, and crest of gentle kind;
> *The Scottish Barnacle, if I might choose,*
> *That of a worme, doth waxe a winged goose.*'

---

* Vide author's 'Seaside and Aquarium.'

CHAPTER IX.

Phyllodoce Laminosa—The Laminated Nereis.

"His meaner works
Are yet his care, and have and interest all—
All, in the universal Father's love."
—COWPER.

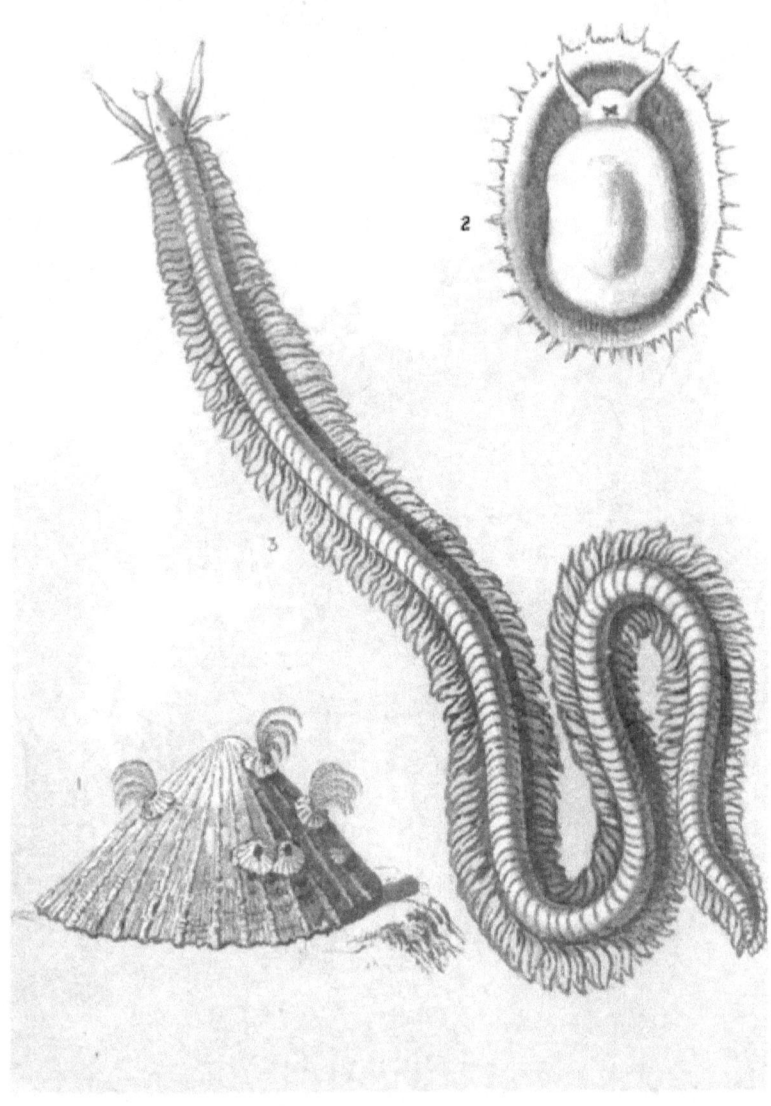

1 COMMON BARNACLES attached to shell of Limpet
2 ANIMAL OF THE LIMPET *(P. vulgata)* as seen from below
3 THE LAMINATED NEREIS *(Phyllodoce laminosa)*

IX.

To oblige an English correspondent who requested some blocks of stone containing Pholas perforations, the writer, in company with a fellow-student, started betimes for the sea-shore, some four miles' distance. We made for a certain spot, where it was expected the object of our wishes could easily be found. Our equipment consisted merely of a hammer, a bottle, and two chisels, enclosed in a carpet-bag, the better to mask our mission from impertinent curiosity.

On reaching the shore, it soon became painfully apparent that no pieces of rock could be procured of a character at all suitable for a museum. To make matters still more irritating, a breeze arose, and with it came a furious shower of rain, which soon completely saturated our light costume. At such a time it is laughable to note how faint becomes the poetry of practical zoology—how excessively like street puddles are the fairy-grots, as the rock-pools are called; how unsightly the great, distorted anemones look, too, when viewed from beneath a large boulder, where you are crouched in the fond idea that you are thus getting shelter from the rain.

On this occasion, my friend and I, being soaked to the skin, started up from our unpleasant position, and boldly daring the rain to do its worst, proceeded to hunt after any object of interest that might by chance be lying stranded near.

After an hour's search, two objects, among many others of more or less interest, were captured, that fully repaid us for our uncomfortable 'ducking.' The first was an elegant Actinia (*A. Dianthus*), which seemed to be exquisitely modelled in the finest virgin wax. The second was a specimen of the Laminated Nereis (*Phyllodoce Laminosa*), a wonderfully beautiful *worm!*

Fair reader, start not nor curl that rosy lip of thine at the expression, 'beautiful,' being applied to such an humble creature, for indeed the title is a just and true one.

This Annelid is generally found coiled up and attached to the under part of stones situated near low water mark. Its general colour is emerald green, excepting along the centre of the back, which is iridescent, and reflects a brilliant blue, changing into purple and other hues, only equalled in beauty by the enamelled corslet of the brightest beetle, or the flashing tints that dance upon the plumage of the humming-bird.

The body of the *P. Laminosa*, like that of all

other *Dorsibranchiate Annelidans*, is divided into a consecutive series of rings. Upon either side of each ring is situated a singular appendage, which acts as a gill or branchial organ, by the exercise of which the blood of the animal is effectually purified, and respiration adequately provided for.

When the Nereis is in a state of repose, these gills are laid flat over its back; but in a state of activity they are fully spread out, and act as 'paddles,' by aid of which the animal is enabled to glide through its native element with a graceful serpentine motion.

At the base of each paddle is situated a smaller one, consisting of a fleshy pedicle shielding a fan-like bunch of hairs, each of which tapers to a sharp point. Combined, these hairs or spines form a powerful defensive weapon, which can be extended or retracted at will; and it also serves as an *oar*, or propeller.

As a noteworthy instance of tenacity of life in the lower animals, it may be well to mention here that I have on various occasions, by aid of the microscope, watched for several minutes the bunch of spines, above alluded to, thrust out and retracted in a single segment cut from the body of the Nereis; and only as the object became devoid of moisture did its beautiful mechanism cease to play.

The specimen now before me is comparatively small, being only twelve inches in length, yet its body contains nearly one thousand lateral appendages, constituting, it must be admitted, a most extensive and wonderful locomotive apparatus.

This Annelid is not a suitable object for the aquarium, on account of its frequent great length, and the consequent likelihood of its getting entangled among stones and rock-work when in search of food.

If the hinder parts be cut off, as has been already hinted, they will exhibit vitality for a considerable period when placed in water, but we are told it is the anterior (?) portion of the Phyllodoce which alone possesses the power of regenerating lost segments; these will be reproduced sometimes at the rate of three or four in a week.

'These creatures,' says a learned author, 'as might be expected from their activity and erratic habits, are carnivorous; and innocent and beautiful as they look, they are furnished with weapons of destruction of a unique and most curious description. The mouth of the Nereis would seem at first to be a simple opening, quite destitute of teeth; but on further examination, this aperture is found to lead into a capacious bag, the walls of which are provided with sharp, horny plates, even more terrible than those which are occasionally to be met with in the gizzards of some of the higher animals. It is not surprising, therefore, that by many anatomists the structure in question has been described as a real gizzard, or by some as the stomach itself. A little attention to the habits of the living Annelid will, however, soon reveal the true character of the organ. No sooner does the creature wish to seize its food than this so-called gizzard is at once turned inside out, in which condition it protrudes from the mouth like a great proboscis, and the teeth, which were before concealed in the interior of the cavity, now become external, and display as formidable an assortment of rasps, files, knives, saws, hooks, or crooked fangs, as any one could wish to see. Let us suppose them, when in this condition, plunged into the body of some poor helpless victim, while at the same moment the proboscis is rapidly inverted and withdrawn; the prey thus seized is at the same instant swallowed, and at once plunged into a gulf where all struggles are unavailing, there to be bruised, and crushed, and sucked at leisure.'

There is a curious fact in connection with these Annelids which is too interesting to be omitted here. I allude to the wonderful manner in which their young are produced by a process that may be called 'sprouting.'

This invariably takes place in the segment immediately preceding the terminal one. When a new animal is about to be formed, the reproductive segment swells, and after a certain time the infant worm is seen growing from the tail of its parent. When sufficiently developed, the offspring detaches itself, and starts life on its own account. Sometimes before the elder born Annelid is fully formed, the mysterious segment produces a second **offspring, and,** according to Professor Milne Edwards, as many as six young ones may be generated in succession from the same posterior segment, all of which will for some time continue attached to the parent worm.

CHAPTER X.

The Fan-Amphitrite.

X.

AT the lowest ebb of spring-tide may often be seen protruding above the surface of the beach an object that at a little distance might be mistaken for the twig of a tree, or a decayed and blackened reed. A close examination discloses it to be a smooth, tough tube, apparently composed of dark leather or old gutta-percha, affixed at its lower extremity to some rock or other solid substance.

The pretty Annelid occupying this dark cylinder is the Fan-Amphitrite (*A. ventilabrum*). Unlike the Terrebella, this animal may really be captured without much difficulty. The first time I made the experiment it was successful. By carefully digging down with chisel, or digits, to the base of the tube, which may be reached in the course of a few minutes, the entire structure, with its living occupant, may be transferred to your extemporaneous tank.

I have an Amphitrite in my aquarium at the present time displaying its richly-tinted tentacula to the sun, which lights them up with unusual beauty.

As the 'case' of this animal is flexible, and as its owner will only thrive in an upright position, the reader will easily conceive that to afford the Annelid suitable accommodation in the aquarium is not a very easy task. What other naturalists do I cannot tell; but the following is the plan I adopt for the creature's comfort and my own gratification:—

Having procured a small cylinder of glass (or gutta-percha), close up one end, and drop in the Amphitrite, taking care to first tie the lower portion of its sheath with a piece of thread or silk. It is very pretty to see the plume of the Annelid spreading completely over and covering the extremity of the tube, giving the idea in the one instance that the animal was mysteriously gifted with the power of exuding gutta-percha instead of its usual mucus.

The Annelid may be made to recline against the sides of the vase, or be propped up on any chosen spot by aid of a small cairn of pebbles, and thus form a very curious feature in the aquarium.

To test a fact, relative to the power which the Amphitrite is said to possess, in common with other tubiculous Annelids, of renewing certain portions of its body after sustaining injury, I snipped off the principal portions of its branchiæ, and found that, after the lapse of a few months, my specimen renewed its mutilated organs.

CHAPTER XI.
The Common Mussel.

'Travelling is not good for us; we travel so seldom. How much more dignified leisure *hath a Mussel glued to his impassable rocky limit two inches square!* He hears the tide roll over him, backwards and forwards, twice a day (as the Salisbury coach goes and returns in eight and forty hours), but knows better than to take an outside place on the top on't. *He is the owl of the sea, Minerva's fish, the fish of wisdom.*'

C. LAMB to B. BARTON.

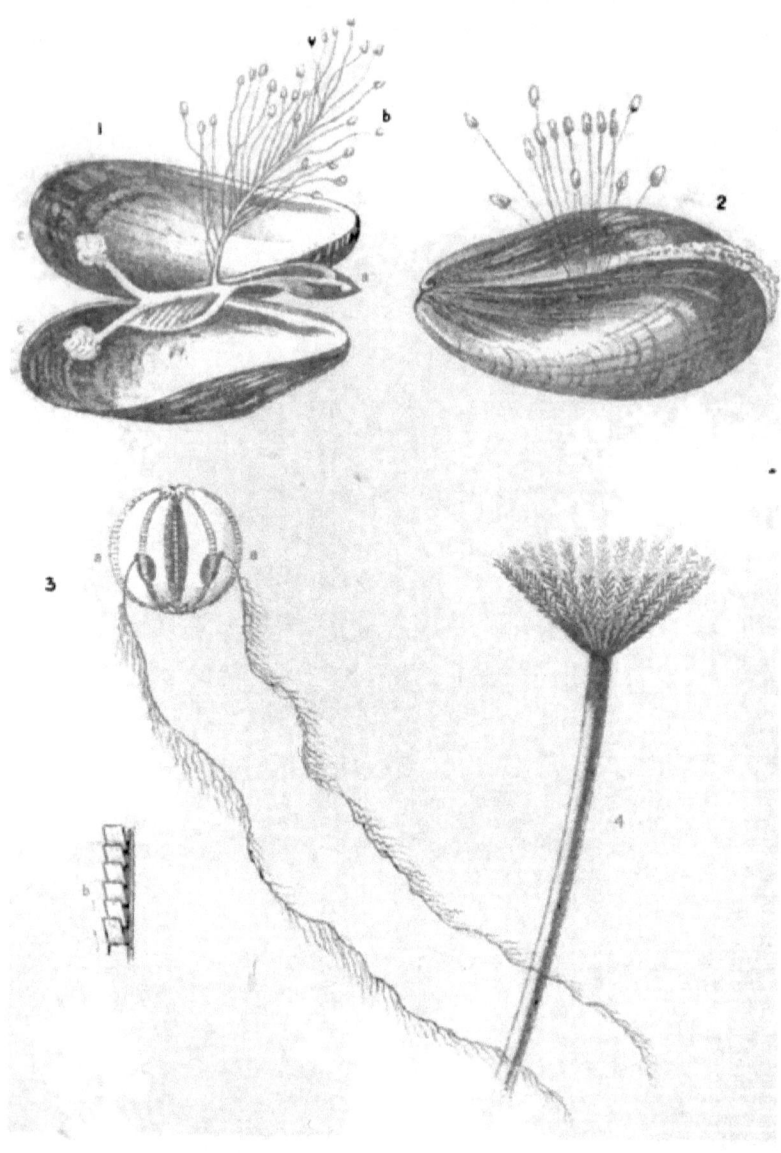

1. COMMON MUSSEL *(Mytilus edulis)*
   a. The foot. b. The byssus. c.c. Muscles which regulate the action of the foot.
2. THE MUSSEL CLOSED.
3. THE BEROE *(Cydippe pileus)*
4. THE FAN AMPHITRITE *(A. ventilabrum)*

XI.

In his celebrated journey to the western islands of Scotland, Dr. Johnson tells us that when at Ulinish, hearing of a cavern by the sea-side remarkable for powerful reverberations of sound, he determined to pay a visit to the spot. After dinner, having procured the services of some boatmen, the doctor, in company with Bozzy, started off on his trip, which, on the whole, appears to have been a pleasant one. There was, however, no *echo* to be heard; but to make up for this disappointment, Mr. Boswell went angling, and caught a wee ' cuddy,' (a fish about the size of a gudgeon), while the doctor was gratified by the sight of some sea-weed growing upon stones, and above all, at witnessing for the first time *Mussels in their natural state.*

The impression made by this candid acknowledgment upon our minds is one of wonder, that a man like Johnson could have reached his advanced years without having seen so common a sight. But it is possible that even in our day, with its unprecedented facilities for cheap travelling for the most inland inhabitants, there may be many persons to whom the sight of a Mussel fixed to a boulder by its self-constructed cable, would be as great a novelty as it was to the eloquent author of Rasselas.

It is, however, one of the commonest appearances which meet the eye of those in the habit of visiting the sea-shore. At certain localities myriads of Mussels may be noticed attached to the surface of the rocks. So thickly are these sometimes covered over, that the blade of a knife cannot be inserted at any part without touching one or more of the esculent bivalves that are to form the subject of this chapter.

The Mussel anchors itself by means of the Byssus; or, as it is commonly termed, the 'beard.' This appendage is composed of various slender threads which are attached to any object within reach, whether such be the shell of a neighbouring Mussel, a small stone, or huge boulder. The members of each colony are thereby bound together, it may be figuratively said, by the silken cords of friendship, and mayhap of love. The *Mytili* evidently believe that 'there's no place like home.' Although gifted with a power of moving about at will, they never attempt to exercise this when living together in a family circle, but pass through life's stages upon the spot where they were born. Certainly, if there be such a sight as a truly happy and contented family in the marine animal kingdom, it is to be found exemplified in these bearded molluscs.

As hinted, they live shoulder to shoulder, back to back, and otherwise mutually support each other. They need not look about for a single meal, but have merely to allow themselves to be fed by the waves, which yield them a constant supply of fresh and wholesome food. Their sole duty in this respect is limited to the selection of objects suitable to their palate. Their 'at homes' being so frequent, the *Mytili* can boast of a large circle of acquaintances. The Periwinkle, and his friend Silver Willie, often make a morning call, take pot luck, as it is termed, and then politely retire. *Mr. Carcinus Mœnas* and his poor and dirty relation, *Maia Squinado*, perchance look in of an evening. *Solaster Papposa*, or occasionally the lanky-legged *Uraster Rubens*, and other 'stars' of the marine world, crawl in at unseasonable hours in their usual lazy style, and are generally rewarded by finding the doors (valves) shut against them. This 'cut direct' does not appear to be at all annoying; or if so, the Star-fishes are too cunning to show it, for they quietly saunter away as if they never had the slightest wish to put their feet within their neighbour's dwelling.

There is a 'black sheep,' as Sir Pertinax Mac Sycophant would say, who intrudes himself into

Mussel society, and plays sad havoc among its members. This crawling rascal is the wolf of all Musseldom flocks. Young and old alike experience the blighting effects of his villanous propensities. The name of this obnoxious personage is *Purpura Lapillus* (Common Whelk). What, the reader will ask in surprise, a univalve prey upon a bivalve? Is that possible? It is, unfortunately, too true.

If we take a Mussel in our hand we shall find it perfectly impossible to force its valves asunder, without the aid of a strong knife or other instrument; yet the Common Whelk, fleshy and insignificant creature though it be, will consume the animal within, and make the valves fly open in a brief space of time, by means of its soft tongue. But leaving such general remarks, let us suppose we are standing before a boulder covered with these mussels. Numbers of gaping shells may be at intervals perceived still attached to the rock, but with the interior of each valve so empty and smooth, that we could scarce believe they had ever embraced a living occupant. On taking up one of the valves and closely examining it, do you observe nothing peculiar about it now? 'No.' Take up the other then, and submit it to a similar inspection. Well, what do you see now? 'Nothing,' you still reply, 'unless it be a peculiar little hole about the size of a pin's head, which surely is of no importance.' That little hole was of *vital* importance to the poor mollusc, for through that aperture the life and substance of the Mytilus was drawn by the voracious Purpura.

But the poor Mussel is exposed to the attacks of other enemies — aquatic birds, as sea-gulls and ducks, eagles, vultures. Even water-rats and monkeys may also be included in the list.

It is amusing to see a gull, by no means a foolish bird, standing patiently before a Limpet, for example. The animal, unsuspicious of the presence of an enemy, raises his canopy with the view of relaxing his overstrained muscles, and is instantly toppled over by the intruding beak of the bird. If unsuccessful in his first attempt, the gull is well aware it would be useless to try a second time at that tide.

But if a Mussel be the object of attack, it is wrenched from its seat, raised to a certain height, and then allowed to drop upon a stone with the view of breaking the shell. In one locality called Mussel Bay, Mr. Barrow says he disturbed some thousands of birds, and found so many thousands of shell-fish scattered over the surface of a heap of shells, that, for aught he knew, would have filled as many thousand waggons.

This habit of the feathered tribe was, by the way, well known to the ancients, and I may be pardoned relieving my pages by a quotation on the subject from the 'Shepherd's Calender' of Spenser, whose exquisite descriptions of natural history are as marvellous as his allegorical poem. The author of the

'Fairy Queen' thus humorously reads a lesson to an ambitious man,—

> "He is a shepherd in gree,
> But hath been long ypent,
> One day he sat upon a hill,
> As now thou wouldst mee;
> But I am taught by Algrinds ill,
> To love the lowe degree.
> For sitting so, with barred scalpe,
> An eagle soared hye,
> *That weening his white head was chalke,*
> *A shell-fish down let flye!*
> *She weened the shell-fish to have broke,*
> *But therewith bruised his brayne,*
> So now astoined with the stroke,
> Hee lyes in lingering payne!"

It seems remarkable that the 'illustrious French naturalist,' Reaumur, should have been the first, if not to discover, at least to publish, any description of the manner in which the Mussel spins its silken cable. Yet one hour's experience in a tea-cup or tumbler will exhibit most of the features in this interesting process.

That Reaumur's narrative, although usually copied by most writers of the present day, is not strictly correct, and, moreover, that the foot of the mussel is *not* 'useless as an instrument of progression' (as generally asserted), may be easily proved to the satisfaction of the student by adopting some such simple experiment as that which I am now about to describe:—

Being at the sea-side on a fine summer afternoon, I procured three specimens (I might have had as many hundreds if disposed) of the Mytilus. On my return home I placed them in a common tumbler, and waited patiently to see the result. My object was, if possible, to witness the manner in which this animal grows its beard.

In less than five minutes an industrious little fellow, whom we will call No. 1, gently opened his shell, and immediately protruded his fleshy foot until it reached a length of nearly two inches. So far as I could determine, the design of the Mussel was to discover, in the first place, what kind of a lodging he occupied; whether or not he had any companions; and also, to know if these or any other objects could be found worthy of his *attachment*.

Sometimes the foot would be protruded under the shell, then in a contrary direction. Or by an exertion of the strong muscular power which that organ possesses, the entire shell would be lifted off the ground and urged forward to a considerable distance. Of course he soon come in contact with a neighbour Mussel, whom we may term No. 2, but as the latter was not anchored by any byssus, he was speedily pushed on, and on, until No. 3 was met, and the latter, in his turn, made to take up a new position.

Being tired of wandering about, No. 1 then extended his foot along the base of the vase to a certain point, and there let it rest for a few seconds. When again withdrawn, to my great delight, I saw the first thread of a new byssus had been constructed.

As my principal object was to become acquainted with the mode of formation of the beard, I did not feel satisfied with merely watching the movements of the animal from above. After a brief interval another thread was spun. I bore in mind the words of Reaumur, who says, 'The Mussel never spins more than four or five threads in the twenty-four hours.' Aware that no time must be lost, though still afraid to disturb the mollusc lest it might suspend its labours, I instantly detached my specimen, and again turned its shell round so as to bring the opening of the valves against the face of the glass. The creature did not seem at all offended at his handiwork having been destroyed, but still obstinately refused to let me see the working of its foot. Again was the shell rolled over, and again did I replace it in its former position. This time, in order to keep it from being shifted, a stone was deposited upon the valve. Nothing daunted, the animal gradually separated the valves of its shell, and at the same time advanced and elevated its foot to the exact position that I had so long desired.

The spinner, when at its full length, was pressed firmly upon the flat surface of the glass, and there allowed to remain for a while. Suddenly, at nearly *half an inch distance from its extreme end* (or point), a little mouth was seen to form, about the size of a large pin's head, from which there issued a milk-white fluid, that gradually hardened and became

CONSTRUCTION OF THE BYSSUS. 173

fixed to the glass. This object being light in colour, had a pretty effect when contrasted with the rich brown tint of the spinner. Shortly afterwards the foot rolled over and withdrew into the shell, leaving behind it the silken thread which had just deen spun. The 'little mouth,' above described, was, if I may so term it, the mould in which the end of the thread was cast.

In the course of two hours a bundle of byssus threads, sixteen in number, were produced by this industrious little labourer.

Having thus seen that the foot is useful to the Mussel as an instrument of progression *before* the beard is formed, let me now endeavour to show that it is, at times, of equal service for the same object, *after*, and when the mollusc is anchored thereby to any particular spot.

When we remember that this anchorage is formed of a harp-like set of strings, amounting to ten or even *a hundred* in number, it does seem an almost incredible fact that the Mytilus is enabled to change its station, even when living in single blessedness.

To see a Mussel 'flit,' is a sight one may often watch and wait for without success. On the other hand, when least expected, the self-willed mollusc may commence operations. When about to take up a new home, the animal shaves off its beard entirely, or in more scientific language, 'rejects its byssus' altogether. In order to excite the locomotive instincts of my specimens, I used to cut all the threads of their cable except one. The animal being suspended, of course its whole weight was then thrown upon a single fibre. Such a state of insecurity was by no means agreeable, and I generally found in the course of a few hours that fresh threads were rapidly thrown out, and an entirely new byssus formed; the old one, which was broken off at the root, being left behind as useless.

Another singular peculiarity of the Mussel which came under my observation has not been, so far as I am aware, noticed by previous naturalists. I allude to the power which the animal possesses of lengthening out the root or stem of the beard, apparently to an unlimited extent. This power appears to be seldom exercised, for although I have had hundreds of specimens of the Mytili, in only one instance have I witnessed the phenomenon in question.

A large specimen of this bivalve, procured accidentally from a fishwife in the street, was dropped into the aquarium, and placed close against the surface of the glass. The animal seemed highly delighted with its change of situation, for in a few moments the valves were opened, and a long draught of water taken in to bathe its branchiæ, and furnish a hearty meal. Having satisfied its appetite, the next process, of course, was to find out what kind of a home he had been introduced into. The foot, a noble specimen, was soon protruded, and one after the other, in rapid succession, various threads were formed. By next morning the animal, advancing by a series of easy stages, had reached the surface of the water, which was exactly five inches deep. I knew it would not remain long in this position, and was anxious to discover what plan would next be adopted. Several courses were open to him. For instance, like a marine Captain Cook, he might circumnavigate his little *Globe*,—or he might let go his cable and drop plump to the bottom,—or he could follow the route I had often seen taken by his relations, viz., to journey back to the place from whence he started. It pleased him, however, to strike out into a new path,—to devise a method of his own. While located near the top of the tank, he threw out exactly ninety-eight threads, not certainly for security, but merely, it would appear, for pleasure.

Then slowly but surely, day by day, he lengthened out the stem of his byssus tree, until it reached the extreme length of nearly five inches. To what further degree it would have been extended, had the mollusc not reached the base of the tank, it is impossible to conjecture.

No sooner did the shell touch terra firma, than the cable which had taken so long to spin was immediately broken off. I have succeeded in keeping the same animal by me for the last twelve months, but have seen no attempt at a renewal of the operation, in the progress of which I had taken so lively an interest. I may add that this Mussel taught me another lesson; it was this: in my early studies regarding the habits of the Mytilus, I had adopted a certain theory of the manner in which the beard was formed; and having watched so long, and witnessed the process so repeatedly, I thought myself justified in forming certain conclusions. My belief was that the creature could not form more than one thread at a time, *without withdrawing its foot into the shell*, as I believed, in order to procure a fresh supply of material. That this notion was erroneous, this animal proved to my entire satisfaction. Not only may one, but two, three, four, and even six threads be attached to any selected object, the point of the foot being passed from one position to another, without the organ being withdrawn into the valves until the whole of the threads are formed. How many more the Mussel is capable of producing at one 'stretch,' I have no means of knowing, but six is the largest number that any of my specimens in such case have ever fabricated.

The general idea seems to be that the Mussel works in the same manner as the spider, who emits a drop of liquid against some foreign substance, which, being allowed to harden somewhat, is then drawn out as the spider recedes. This notion, I may state, is quite erroneous. When the sucker of which we have spoken is formed, *the thread is completed*. It is true that the foot as it retires into the shell generally glides down the newly-constructed filament, but this is not of necessity, nor does such circumstance invariably occur. Indeed, while busily engaged in attaching a disc to the glass, the muscles of the foot will contract, and thus throw open the folds of the groove, situated in the middle of that organ; when thus exposed, the byssus thread may be seen in the furrow, stretched like the string of a harp or dulcimer.

While the end of the thread is being attached to a certain spot, a conspicuous muscular action is perceived going on in the foot, which alternately swells and contracts, as if something were being pumped up through the byssal channel, until it reached a certain point. There being dilated and spread out in successive layers, it assumes a trumpet-like disc, which is firmly fixed to the foreign object. Indeed, I am by no means certain that the thread is not, when first produced, exactly like a trumpet in shape. It also conveys the idea of being blown out in a similar manner to a piece of bottle glass. After being exposed to the air for some little time, the hollowness of the thread is not so apparent as when it is newly fabricated.

The mucous fluid, from which the fibres are formed, is secreted in a gland situated at the base of the foot, whence it is apparently expelled at the will of the animal into the furrow already referred to, and is there spun into threads. The toughness of these filaments, considering that each is finer than the thinnest strand of silk, is remarkable. Their strength, however, may be easily accounted for, when we know that each is composed in reality of innumerable delicate threads, bound together by a subtle gelatinous fluid. This phenomenon may be made out quite distinctly with a common hand lens, if the following simple experiment be adopted: Make a Mussel construct its thread in such a way that the disc of each is planted on the face of the glass. Then place the fine point of a common needle upon the outer edge of a chosen disc or sucker, and gently draw the former away to a little distance, and you will find that by so doing the stretched string becomes *peeled*. Continue this process carefully, and before the thread gives way you will have divided it into a dozen parts at least, all of which are visible to the naked eye, but clearer when the hand lens is used, and still more distinctly and beautifully defined, of course, if the microscope be brought into play.

The foot of the Mussel appears to be firmly strapped on, as it were, to certain transverse muscles, by a contraction of which the animal closes its shell with surprising force. This strap, composed of a powerful tendon which passes under the adductor muscles, is attached at either end to the base of the foot. Thus we account for the remarkable strength which is evidently seated in the foot, and makes it of so much importance to the animal. At first sight nothing appears more easy than to pluck out this organ by the roots, but an attempt will prove the experiment to be more difficult than many persons suppose.

The colour of the foot varies considerably in different specimens, even of the same species. Some, for instance, are of a chesnut brown; others of a kind of mauve or purple, covered with a peach-like bloom during life; others, again, are of a deep-toned umber, while not a few are pearly white, and streaked sometimes with pink like a tulip.

The peculiarity of the Mussel to attach itself to foreign substances has been taken advantage of for the benefit of man, and a curious instance is exhibited at Bideford in Devonshire, at which town there is a bridge of twenty-four arches, stretching across the Torridge river near its junction with the Taw. 'At this bridge the tide flows so rapidly that it cannot be kept in repair by mortar. The corporation, therefore, keep boats in employ to bring mussels to it, and the interstices of the bridge are filled by hand with these mussels. It is supported from being driven away entirely by the strong threads these mussels fix to the stonework.'

Like most other writers who quote this strange account, I have not had ocular proof of its accuracy.[1] That it is quite probable I can readily believe, as a

---

[1] Since writing the above, I have received the following interesting epistle from Mr. Edward Capern, the celebrated 'poet and rural postman' of Bideford, who kindly

pretty experiment will partly prove it to any spirited aquarian. Following out the above idea of the bridge at Bideford, I managed to build an exceedingly pretty centre piece for my tank.

Having no ready means of making a rock arch, I collected such pieces of rock, stones, &c., with weeds attached, as I thought would answer my purpose, and then proceeded to fabricate the object of my wishes in the following simple way: First were laid two stones parallel to each other at three or four inches apart. Upon these I placed a large piece of rock in a transverse direction. Between the interstices a number of small mussels were then inserted. When fully satisfied that the bivalves had moored themselves, I gradually piled one piece of rock upon another until the structure reached the desired height, each piece being bound to its neighbour by means of the byssus threads of the Mytili.

Before each block of stone that formed the foundations of the arch was placed a splendid frond of Lettuce Ulva, tied by a strand of silk to a white pebble. These verdant fronds, so smooth in texture and so gracefully convoluted, rising up from the base

---

sought out the information I desired, relative to the present state of the bridge above alluded to :—

Bideford, January 27, 1859.

DEAR SIR,—I have inquired of the bridge warder, and he informs me that the feoffees of the bridge command mussels to be brought up by the cart-load, to protect the foundation, which is laid on *rubble*.

I am pleased that it has been in my power to procure this information for you —I am. dear sir, faithfully yours, EDWARD CAPERN.

of the tank and reaching to its brim,—mingling, too, with the various tufts of corallines and other seaweeds that jutted from each crevice, were very pretty to look at. When disturbed by the movements of the fishes passing in and out, the gracefulness and beauty of the sea-weed was doubly increased.

In making observations upon any bivalve, such as the Mussel, it is extremely puzzling to know what is going on *inside* the shell. Yet it is almost necessary to acquire this knowledge by means not always apparent, in order to satisfy one's mind relative to certain appearances, which we perceive going on externally. We have to form our judgment of things we do not see from those that are apparent—at all times a difficult task. But not often so tantalizing as in the case of an insignificant creature like the Mussel, who lives, moves, and works constantly before our eyes. I may add that it was not enough for me that I saw the spinning process frequently. It all seemed tolerably clear to my mind, but still I did not feel thoroughly satisfied. My desire was to peep into the shell, and find out where the last spun thread was situated; or, in other words, from what part of the trunk the new branch sprung. On examining various specimens of the byssus, this point was by no means apparent. Various means I adopted failed to secure me the requisite knowledge. At length I hit upon a plan, which, after no long time, I found opportuuity to put in practice. My largest

Mussel lifted up its testaceous canopy, put aside the fringed and fleshy veil that surrounded its edge, protruded its spinner to make sure the ground was secure, and then withdrew it again into the shell as usual. After the lapse of a second, the foot re-appeared and was stretched out to an unusual length. No sooner was the end of the thread formed on the glass than immediately I firmly pressed the valves together, and held them in this position until I had gradually worked the Mussel up out of the vase, when I bound them close together by means of a piece of cord. I need not describe my manœuvres further; suffice it to say that the thread nearest to the base of the grove was found to be the one that was spun last. This, in my opinion, is invariably the case.

I may mention that the above experiment also proved to my mind that the foot must be a most important vital organ of the Mytilus. At times, on placing an open Mussel in my tank as food for crabs or other animals, I have noted that if every other part were eaten, and the foot allowed to remain attached to the muscles of the bivalve, that member would after a lapse of several days show signs of—I do not say life—but sensation and retractile power.

But when the foot is cut and otherwise injured, the animal dies quickly. In the experiment mentioned the valves were not kept closed for more than half an hour; yet when they were opened, vitality had evidently ceased within. This was the more singular when we remember that the Mytili will live for many days out of the water; the shells, of course, during the whole period being firmly closed.

The Mussel, as already hinted, is very tenacious of life. I have kept specimens by accident for several days in the pocket of my coat, but found them quite well and lively when placed in sea-water.

In general the sure sign of their not being in a healthy condition is when the shell opens; for, while the animal retains any sense whatever, it exercises a strict and judicious 'closeness.'

I have found, however, on several occasions, that the shell being contracted is not always a valid proof of its owner's convalescence, for when placed in water the Mussel would float for several days upon the surface like a cork, although it was near death's door.

This phenomenon must be caused, I should suppose, by some sudden fright compelling the mollusc to close its shell with such rapidity as to prevent a proper supply of water being taken in. Having only air to exist upon, the animal then lingers on until its branchiæ become dried up, and all moisture exhausted. In this state the Mussel opens its shell with a deep bursting sigh, and sinks to the bottom—dead.

Being at the sea-side one fine summer day, I heard a little Scotch girl cry out to her brother who was about to swallow entire, a fine specimen of the

Mytilus edulis, ' Eh, Willie dear, dinna ye eat that. Dinna eat the *beard* or ye'll dee !' Many years ago I remember a remark to the same purport as the above being made by a poor child to its playmate, in the neighbourhood of Gravesend.

I little thought at that time that the Mussel was so interesting a shell-fish, or that I years after should spend many an anxious hour studying the formation and nature of its despised beard.

I need hardly state that the idea of the beard being poisonous is a vulgar error. In general the fish may be eaten entire with impunity.

Cases have occurred where persons have been taken ill after eating it, but this result has been satisfactorily explained to have been caused by the Mussels being procured from places such as Leith Docks, where their food consisted chiefly of unwholesome and putrescent matters.

This mollusc is not used as food to any very great extent by the poorer classes. It is employed very extensively, however, by the fishermen as bait along all parts of the British coast. But in France it is much esteemed both by rich and poor. The trade in them is successfully cultivated, and affords a means of support to hundreds of industrious and deserving men.

From the learned author of the " Rambles of a Naturalist" we learn that at the village of Esnandes, on the coast of France, the Mussel trade, commenced about eight hundred years ago, has assumed a gigantic extent. Both here and at the neighbouring villages of Charron, Marsilly, Mussels are bred in an ingenious and systematic manner. At the level of the lowest tide short piles or stakes are driven into the mud, in a series of rows about a yard apart. This palisade is then roughly fenced in with long branches. On this structure the Mussel spawn is deposited, and it is found that the molluscs thus produced in the open sea are much finer than those which are bred nearer the shore.

These artificial Mussel beds are termed 'bouchots. The fishermen who engage in this branch of industry are known as ' boucholeurs.'

'The little Mussels,' continues M. Quatrefage, ' that appear in the spring are known as *seeds*. They are scarcely larger than lentils, till towards the end of May, but at this time they rapidly increase, and in July they attain the size of a haricot bean. They then take the name of *renouvelains*, and are fit for transplanting. For this purpose they are detached from those *bouchots*, which are situated at the lowest tide mark, and are then introduced into the pockets or bags made of old nets, which are placed upon the fences that are not quite so far advanced into the sea. The young Mussels spread themselves all round the pockets, fixing themselves by means of those filaments which naturalists designate by the name of byssus. In proportion as they grow and become crowded together within the pockets, they are cleared out and distributed over other poles lying somewhat nearer to the shore, whilst the full-grown Mussels which are fit for sale are planted on the *bouchots* nearest the shore. It is from this part of the Mussel bed that the fishermen reap their harvest, and every day enormous quantities of freshly gathered Mussels are transported in carts or on the backs of horses to La Rochelle and other places, from whence they are sent as far as Tours, Linoges, and Bordeaux. . . . . The following data, which were collected by M. D. Orbigny more than twenty years ago, will show how important this branch of industry must be to the district in which it is cultivated. In 1834 the three communes of Esnandes, Charron, and Marsilly, representing a population of 3000 souls, possessed 340 *bouchots*, the original cost of which was valued by M. D. Orbigny at 696,660 francs; the annual expenses of maintaining them amounted to 386,240 francs, including the interest of the capital employed, and the cost of labour, which, however, is spared to the proprietor who works on his own account. The nett revenue is estimated at 364 francs for each *bouchot*, or 123,760 francs for the three communes. Finally, the expense of the carts, horses, and boats, employed in transporting the Mussels, then amounted annually to 510,000 francs; but these numbers are far from representing the expenses or profits at the present day. At the time M. D. Orbigny lived at Esnandes, the *bouchots* were only arranged in four rows; now however, there are *no less than seven rows, and some of them measure more than* 1000 *yards from the base to the summit.* The whole of these *bouchots,* which were at first limited to the immediate neighbourhood of the three villages, of which I have already spoken, **extend** at the present day uninterruptedly from Marsilly far beyond Charron, and *form a gigantic stockade for two miles and a half in breadth, and six miles in length.*'

A curious circumstance connected with the Mytilus remains to be described. Let the reader, who may be so fortunate as to possess a good microscope, cut away a portion of the fleshy part of the Mussel, then place it in a watch glass, and examine it through that 'portal to things invisible,' and, unless I am much mistaken, he will own the sight to be supremely wonderful. Some water being deposited in the glass the fleshy object will be seen to swim about in a most singular and mysterious manner, while a close inspection shows every portion of it to be in active motion.

This motive power is caused by countless cilia, the rapid vibration of which creates constant currents. This action preserves the health of the poor mollusc by aerating the water which passes over his respiratory organs.

That some such wonderful contrivance is adopted, for conveying food within the valves, too, is evident, when we consider that the Mussel is always affixed to some foreign substance, that it cannot hunt after prey, and therefore can subsist only upon whatever nutritious particles may be contained in the element in which it lives. These consist of minute animalculæ, principally crustacea, which are drawn within the shell by powerful currents.

I have often watched this phenomenon through a hand lens, and have seen the young shrimps and skip-jacks, for instance, notwithstanding the nimbleness of their movements, irresistibly drawn into the gulf of destruction. Even tolerably sized specimens that were seated in fancied security upon a valve of the Mussel, have suddenly been drawn in, out of sight. As an instance of the power of these currents, I may state that the water in a small aquarium is often seen to be affected by the respiratory action of a single bivalve. The same thing has even been apparent to the writer, while watching the movements of a colony of Barnacles attached to a Limpet, the most distant part of the fluid being gradually drawn near, in obedience to the beck of these delicate and graceful little creatures.

CHAPTER XII.

Terebella Figulus.

(THE POTTER.)

'Whether progressing on the solid surface, or moving through the water, or tunneling the sand, advancing or retreating in its tube, the Annelid performs muscular feats distinguished at once for their complexity and harmony. In grace of form the little worm excels the serpent. In regularity of march, the thousand-footed Nereid outrivals the Centipede. The leaf-armed Phyllodoce swims with greater beauty of mechanism than the fish; and the vulgar earthworm shames the mole in the exactitude and skill of its subterranean operations. Why, then, should the "humble worm" have remained so long without a historian? Is the care, the wisdom, the love, the paternal solicitude of the Almighty not legible in the surpassing organism, the ingenious architectures, the individual and social habits, the adaptation of structure to the physical conditions of existence of these "degraded beings?" Do not their habitations display His care, their instincts His wisdom, their *merriment* His love, their vast specific diversities His solicitous and inscrutable Providence.'—DR. WILLIAMS.

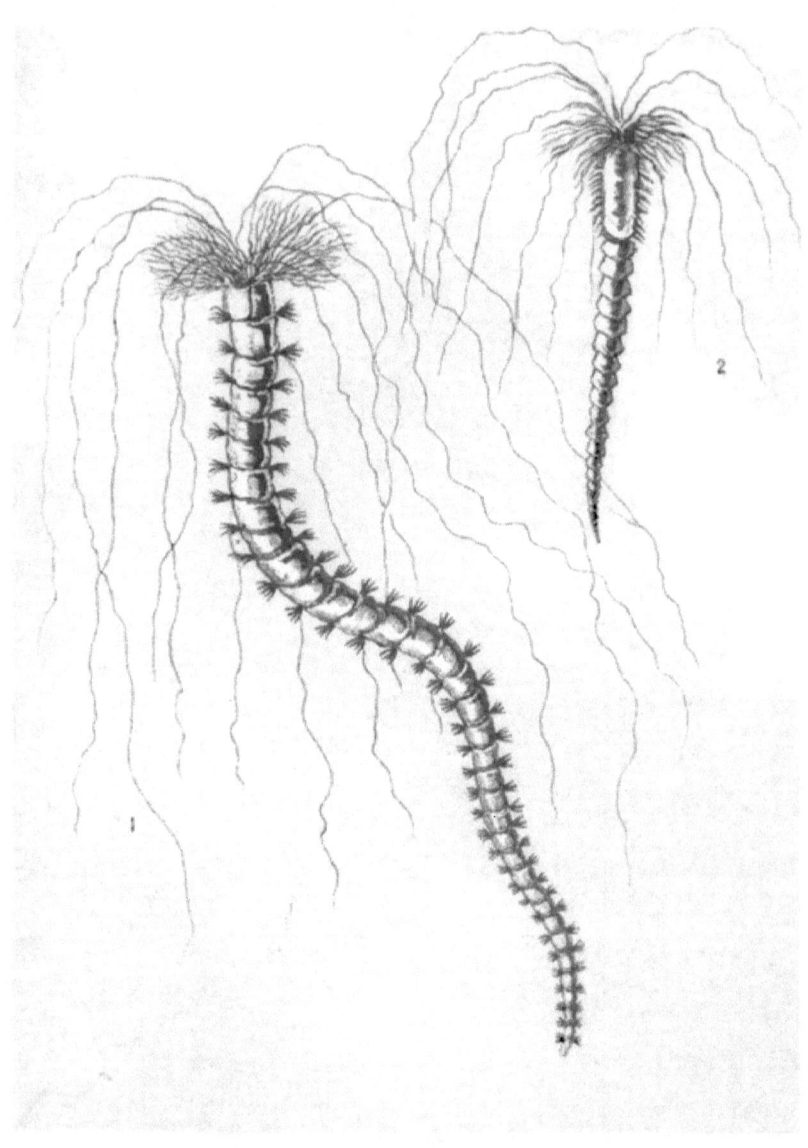

1. THE POTTER *(Terrebella Figulus)*
2. *Terrebella littoralis*

XII.

HAVING visited the sea-side a few weeks since, along with some 'aquarian naturalists,' among other objects we managed to capture a very fine *Terebella Figulus*, commonly called the 'Potter.' The specimen measured about five inches in length, and was nearly as thick as a common drawing pencil. I could discover no signs whatever of any tube in the rocky basin in which the Annelid was situated, a circumstance that struck me as being rather curious.

On returning home, my much-valued prize was placed in a tumbler of large dimensions, the base of which I strewed with newly-pounded shells and gravel. By the following morning all the fine or powdered portion of the 'Silver Willies' had been collected and used in the construction of a tube, sufficient in length to cover half the contracted body of the industrious little mason. After labouring for a fortnight, the tube was gradually extended across the bottom of the vase in a cylindrical form, but eventually it assumed a semi-circular shape, being built upon the glass, and elevated by gentle stages up each side of the vessel until the level of the water was reached, when all further labours ceased for a time.

After the lapse of a few days the building operation was resumed, and the tube carried fully an inch further, at right angles to its former position. The opposite end of the structure was next extended at an angle of 45° from the base of the vase, to a height of about two inches. Then commenced a very curious phenomenon. Some of the tentacles were incessantly elevated and extended across the vessel, until they touched the opposite end of the tube, with what object I could not then conceive. The design, however, was afterwards made evident: in less than two days the animal succeeded *in making both ends of its tube meet together, so as to form a continuous circle.* I happened to be watching the mason, when the last stroke of his labial trowel was given to the building, and shall never forget the cautious way in which the animal crept for the first time over the newly-completed portion of the work, and the seeming delight with which it continued to glide, hour after hour, over the entire circuit of its dwelling-place.

Sometimes its body would be long drawn out, until the tips of the tentacula would reach, and apparently tickle the extreme point of the tail; then a race would commence, in character exactly resembling that so often witnessed with the kitten, or the playful whelp, when either of these animals foolishly imagine that the tip of their tail is adorned with some coveted tit-bit.

The branchial organs of my specimen were very beautiful objects, being formed of three blood-red spiral tufts, the effect of which were heightened by their being placed in contact with the drab, coloured cephalic* tentacles, which seemed to be almost innumerable. These latter organs, although apparently so useless when seen closed, are in reality of the greatest importance to the *Terebella*, for they not only act as auxiliary organs of respiration, and aid most materially in building its dwelling-place, but also constitute the real organs of locomotion.

'They consist,' says Dr. Williams, 'of hollow flattened, tubular filaments furnished with strong muscular parietes. The band may be rolled longitudinally into a cylindrical form, so as to enclose a hollow cylindrical space, if the two edges of the band meet, or a semi-cylindrical space if they only meet imperfectly. This inimitable mechanism enables each filament to take up and firmly grasp *at any point of its length* a molecule of sand, or, if placed in a linear series, a row of molecules. But so perfect is the disposition of the muscular fibres at the extreme free end of each filament, that it is gifted

---

* Cephalic, belonging to the head.

with the twofold power of acting on the sucking and muscular principle.

'When the tentacle is about to seize an object, the extremity is drawn in, in consequence of the sudden reflux of fluid in the hollow interior. By this movement a cup-shaped cavity is formed, in which the object is securely held by atmospheric pressure. This power is, however, immediately aided by the contraction of the circular muscular fibres. Such, then, are the marvellous instruments by which these peaceful worms construct their habitation, and probably sweep their vicinity for food.'

The foregoing beautifully and accurately describes the tentacular cirri. The use of these organs in the formation of the tube in which the Annelid dwells, I will now endeavour to make clear, from close personal observation.

It is an extremely interesting sight to watch a Terebella extend its tentacles in all directions in search of building materials, catching up the surrounding molecules (sometimes visible along the whole length of each filament), and then, by a strong muscular contraction, bringing the collected atoms to the opening of the tube, around which, as is generally supposed, they are then immediately attached by a secretion which is exuded from the body of the animal. Such, however, is not the case.

When the filaments bring their 'subscriptions,' the material thus gathered, instead of being used at once for building purposes, *is, in reality, first eaten by the animal, and, after undergoing a kind of mastication, becomes coated with a salivary secretion, and is then ejected in mouthfuls at the extremity of the tube*, which, by such means, becomes gradually elongated. The shell work, when deposited as above mentioned, is held in position, and prevented from falling over the outside of the cavity, by the filaments which are made to hang down in a most ingenious manner; the animal, at the same time, putting itself in motion, allows the slimy surface of its body to press and rub against the new addition to the tube, which is thus effectually strengthened and soldered together.

The animal does not always wait until the opening of the tube is reached, but gently disgorging while lying at its ease, it then pushes forward by aid of its head and tentacles the mass of building material, which soon becomes distributed and moulded to its proper shape.

If your specimen should happen to build a complete tube, its mode of working cannot be well seen; but should it economize its labours, and run its house up in a semi-circular form against the transparent side of the vessel, as the animal I write of did, you will be enabled to see distinctly every movement that goes on in the interior.

It seems somewhat singular that the Terebella should possess the power of turning itself within its tube, so as to be able to extend its habitation from either end at will. I have very often watched the operation with emotions of pleasure, not unmixed with wonder.

Wishing to test the powers and intelligence of my specimen, I dropped within its tube, which was curved in shape lengthways, some particles of sand, and a pebble which nearly filled up the 'bore' of the cylinder. The great annoyance occasioned by this intrusion to the master of the house was painfully evident. For a whole day the Terebella endeavoured to push out the objectionable matter by means of its head and cephalic tentacula, but without success; for although the mass frequently neared, it never touched or toppled over the mouth of the aperture, and consequently fell to its original position at the bottom of the tube as soon as the animal removed the pressure.

Apparently despairing of its efforts, though still oftener repeated, being eventually prosperous, the poor Annelid literally 'turned tail,' and very coolly proceeded to elongate the opposite end of its dwelling. This operation did not last long; for in the course of a few hours, on peering again into the vessel, I saw that the humble and insignificant worm had mustered up courage to 'face the enemy' once more, and had, in fact, apparently conceived a new idea, the wisdom of which was soon made palpable; for slowly, but surely, most of the arenaceous particles were eaten, and nothing being left but the pebble, it was speedily and triumphantly ejected from the tube, and the sand soon after employed for building purposes. The Terebella having completed its laborious and well-executed task, seemed to be quite exhausted, and lay to all appearance lifeless for the succeeding four and twenty hours.

When walking along the sandy beach, myriads of peculiar objects may be seen swayed to and fro by the roll of the waves. Frequently, when the tide has receded, these tubes remain sticking out of the sand to the height of two or three inches, each terminating in a tuft, like the end of a piece of cord that had been teazed out.

Within such a fragile habitation dwells the *Terebella littoralis*, the most common species of the marine tubiculous Annelids. I have very often tried, by aid of my fingers only, or a strong spade, to capture one of these creatures, but have never been successful, even in a solitary instance. Yet several authors tell their readers the task is perfectly easy. Mr. Lewes, for instance, made me feel ashamed of my previous manipulative efforts when I read his vivid description of a Terebella hunt, and caused me lately to journey a distance of six miles to try my hand again, with no better result than hitherto.

I have lately seen a specimen of *T. littoralis* which a friend of mine was so fortunate as to capture. The tube of the animal, instead of being in its usual position, was situated in a pool, and offered a most rare prize to its discoverer, it being the only one he had ever caught. The tube, being of great length, was cut down to about six inches and transferred to a shallow glass tank, in which was introduced some pounded shells. The beautiful Annelid soon made itself at home, and commenced to repair the damage done to its habitation by collecting these particles, by means of its tentacula, which were thrown out to an extraordinary distance in all directions.

The result of the animal's labours was soon apparent by a most amusing white patch being added to each end of its dark tube.

As soon as this operation was completed *mon ami* carefully tore up the patched garment, and ejected the defenceless Terebella into the vessel, wherein was placed a piece of glass tube that measured an inch in length. Strange to state, the animal instantly crept into this object, and soon made itself quite at home and comfortable. When the building materials were placed near, they were collected and *attached to each end of the glass cylinder* by the little architect, who doubtless was the first of its 'family' who could boast of such a noble mansion,—which ultimately assumed a very remarkable aspect from the variously-coloured 'mortar' that was employed in its construction. Above and below the transparent centre came patches of red, white, and blue material, composed respectively of broken tile, pounded shells, and coloured glass. Such a 'concourse of atoms' was surely never before combined, either 'fortuitously' or otherwise, in the construction of so common an object as the tube of an Annelid.

The branchiæ of the above mentioned specimen presented a most exquisite appearance, resembling the perfect skeleton of a leaf, supposing that to be dyed a brilliant crimson colour, and made to exhibit incessant life-like motion even in its most delicate and minute ramifications.

CHAPTER XIII.

Acalephæ.

(MEDUSÆ, OR JELLY-FISH.)

"And now your view upon the ocean turn,
And there the splendour of the waves discern;
Cast but a stone, or strike them with an oar,
And you shall flames within the deep explore;
Or scoop the stream phosphoric as you stand,
And the cold flames shall flash along your hand,
When lost in wonder, you shall walk and gaze
On weeds that sparkle, and on waves that blaze.'

XIII.

There are certain narrow-minded persons who raise objections to men of science prying into the secrets of nature, and profanely, as they think, attempting to explain the design and purpose of the great Creator.

But to the intelligent and right thinking man, no employment could be found more elevating or ennobling than this; and whether he be a fellow-worker himself, or merely an approving observer of the labours of others, still he feels, and conscientiously believes in the words of Milton, that—

> "The desire which tends to know
> The works of God, thereby to glorify
> The great Workmaster, leads to no excess
> That merits blame, but rather merits praise
> The more it seems excess."

When such a one contemplates the atmosphere, for instance, with its 'wonderful phenomena of clouds, rain, and sunshine, that alternately shield, moisten, and warm the face of the earth, he feels awed by the grandeur of the exquisite system of machinery by which such beautiful results are accomplished. To him also the sea, with its physical geography, becomes as the main-spring of a watch; its waters, and its currents, and its salts, and its inhabitants with their adaptations, as balance wheels, cogs, and pinions, and jewels. Thus he perceives that they too are according to design; that they are the expression of one thought, a unity with harmonies, which one intelligence only could utter.' To his eye all created things possess an interest doubly great, not only from their marvellous structure, but from the mission they are destined to fulfil in this lower world.

What peculiar mission the Acalephæ (which we are now about to consider) were destined to fulfil it has long puzzled men of science to explain. Nor can this be wondered at, when we remember the amazing number of these creatures, and also the extreme delicacy of their structure. Some indeed appear almost as if they were formed by the sportive combination of air and water, as if the sea-breeze ruffling the face of ocean caused bubbles innumerable to arise, which becoming mysteriously endowed with life, thenceforth existed as Medusæ.

They have, indeed, frequently been spoken of as 'animated sea-water,' or 'living jelly.' These expressions seem most appropriate when we remember, that if one of these creatures be placed upon a plate of glass, and allowed to remain exposed to the sun's rays, the only thing that will remain to testify to the existence of this singularly graceful object is a thin film, that a stroke of the sponge or finger will remove in an instant.

The most satisfactory explanation that has been offered as to the use and purpose of the Medusæ is, that *they serve as the principal food of whales and other Cetacea.* To these marine monsters—frequently found from 70 to 110 feet long—we can imagine a few hundreds of jelly-fish would be considered a small meal. The supply, however, is ever equal to the demand, as we shall see hereafter.

I may here be permitted to explain that, in most large fishes, the jaws are completely filled with formidable teeth, as in the shark, for instance. This rapacious monster—which has been aptly termed the tiger of the sea by us, and which the French, in allusion to the deadly character of its habits, have named Requin, or Requiem, the rest or stillness of death—possesses a most marvellous dental apparatus.

Its teeth are not, as might be supposed, fixed in sockets, but attached to a cartilaginous membrane. The teeth, in fact, are placed one behind the other in a series of rows; the first of which, composed of triangular cutting teeth, stands erect and ready for use. But as the membrane continues to grow and advance forward, it slowly perishes, and the teeth drop off, their place being taken by the next row which formerly stood second. These, in the course of time, are succeeded by a third series, which are again followed by others.

Now, whales possess no such weapons. Their enormous mouths are not filled with 'tusks or grinders, but fitted instead with vast numbers of oblique laminæ of a softer substance, usually denominated whalebone, which is admirably adapted for the crushing and masticating of soft bodies.'

To give an idea of the amazing extent of the harvests of 'whale food,' as the Medusæ are termed, that abound in various parts of the ocean, we need only quote the evidence of various navigators on the subject. One (Lieut. Maury), for example, states, that on the coast of Florida he met with a shoal of these animals, that covered the sea for many leagues, through which his vessel, bound for England, was five or six days in passing. The most singular part of the story is that, on his return some sixty days after, he fell in with the same shoal off the Western Islands, and here again he was three or four days in getting clear of them.

The Western Islands here mentioned are, it seems, the great resort for whales; and 'at first there is something curious to us in the idea that the Gulf of Mexico is the harvest field, and the Gulf Stream the gleaner which collects the fruitage planted there, and conveys it thousands of miles off to the living whales at sea. But, perhaps, perfectly in unison is it with the kind and providential care of that great, good

Being who feeds the young ravens when they cry, and caters for the sparrow.'

But Dr. Scoresby, in his work on the Arctic Regions, by aid of figures conveys the most vivid idea of the myriads of these creatures that float in the bosom of the ocean. This writer discovered that the olive-green colour of the waters of the Greenland sea was caused by the multitudes of jelly-fish contained therein. On examination he found that 'they were about one-fourth of an inch asunder. In this proportion a cubic inch of water must contain 64; a cubic foot, 110,592; a cubic fathom, 23,887,872; and a cubical mile, 23,888,000,000,000,000! From soundings made in the situation where these animals were found, it is probable the sea is upwards of a mile in depth; but whether these substances occupy the whole depth is uncertain. Provided, however, the depth which they extend be but 250 fathoms, the above immense number of one species may occur in a space of two miles square. It may give a better conception of the amount of Medusæ in this extent, if we calculate the length of time that would be requisite with a certain number of persons for counting this number. Allowing that one person could count 1,000,000 in seven days, which is barely possible, it would have required that 80,000 persons should have started at the creation of the world to complete the enumeration at the present time! What a prodigious idea this fact gives of the immensity of creation, and of the bounty of Divine Providence, in furnishing such a profusion of life in a region so remote from the habitations of man. But if the number of animals be so great in a space of two miles square, what must be the amount requisite for the discolouration of the sea through an extent of perhaps 20,000, or 30,000 square miles.'

These creatures may be appropriately termed the glow-worms of the ocean, for it is to them that the phosphorescence of the sea is mainly attributable.

Sir Walter Scott, in his poem of the 'Lord of the Isles,' thus alludes to this phenomenon:—

> 'Awaked before the rushing prow,
> The mimic fires of ocean glow,
> Those lightnings of the wave.
> Wild sparkles crest the broken tides.
> nd, flashing round the vessel's sides,
> With elfish lustre lave;
> While far behind their livid light
> To the dark billows of the night
> A gloomy splendour gave.'

Hugh Miller also gives a beautiful prose description of the luminosity of our own seas, but we must resist the temptation to introduce it here.

The appearance of the Greenland Seas is principally owing to the presence of the minute species of Acalephæ, but there are many others that grow to an immense size. Specimens of these may be frequently seen cast on the sea-beach by the force of the waves. When in their native element they form the swimmer's dread, owing to a peculiar stinging power which they possess.

The Medusæ have been divided into groups, and distinguished according to their different organs of locomotion. The common idea is that all jelly-fishes are like mushrooms or miniature umbrellas. Such, it is true, is their general form, but others abound both in our own and in foreign seas, that possess a totally different appearance. For instance, some move by means of numerous cilia, or minute hairs that are attached to various parts of their bodies. By the exercise of these organs the creatures glide through the water, and hence they are called *ciliograde Acalephœ.*

One of the most remarkable examples of this class is seen in the Girdle of Venus (*Cestum veneris*). 'This creature is a large, flat, gelatinous riband, the margins of which are fringed with innumerable cilia, tinted with most lively irridescent colours during the day, and emitting in the dark a phosphorescent light of great brilliancy. In this animal, too, which sometimes attains the length of five or six feet, canals may be traced running beneath each of the ciliated margins.'

This animal, as it glides rapidly along, has the appearance of an undulating riband of flame. Most likely it is the species to which Coleridge alludes in the following passage:—

'Beyond the shadow of the ship
I watched the water snakes
They moved in tracks of shining white,
And when they reared, the elfish light
Fell off in heavy flakes.

> Within the shadow of the ship
> I watched their rich attire—
> Blue, glossy green, and velvet black,
> They curled and swam; and every track
> Was a flash of golden fire.'

Another of this class is the common Beroë (*Cydippe pileus*); its body is melon-shaped, and covered over by rows or bands of cilia, placed similarly to the treads on a water wheel, one above another. These are entirely under the will of the little gelatine. It can use each or all of them, and thus row itself along at pleasure. But perhaps the most singular portion of this creature is what has been termed its fishing apparatus, though by some writers it is considered merely to be the means by which the Beroë anchors its body to any desired spot. It consists of two exceedingly slender filaments or streamers, which measure many times the length of the Beroë itself. Some writers, again, fancy that these organs are used to propel the animal. This must be an erroneous notion, for if they were cut off, the creature would still continue to move with the same power as before. Nay more, if the little Cydippe be cut into pieces, and the ciliated bands be attached to each fragment, the latter will swim about with the same power as when connected with the entire animal.

From the filaments here described, others more slender still depend at regular intervals, which curl up like vine tendrils upon the principal stem. The whole can be spontaneously elongated or slowly withdrawn within the body of the Beroë, where they lie enclosed in two sheaths until again required for use.

These interior 'sheaths,' which resemble in shape the drone of a bag-pipe, are easily seen, being almost the only parts which are not perfectly transparent. They are whitish in colour, and semi-opaque. (Plate VI. contains a sketch of the Beroë, drawn from nature.)

I may mention that the paddles, with their comb-like array of cilia, flap successively in regular order from the top to the bottom of each row. This wave-like movement takes place simultaneously in all the rows, when the animal is in full vigour.

The organs of progression in the *Pulmonigrade* Acalephæ, as their name imports, bear certain resemblance to the lungs in respiration. They move by the expansion and contraction of their umbrella-shaped bodies. Graceful and elegant indeed are the motions of these creatures. I have seen small specimens about the size of a sixpence, advance, in three springs, from the bottom to the top of a large vase in which they were confined.

In descending they turn over and allow themselves to sink gradually as if by their own weight.

The third division of the Acalephæ is termed *Physograde*. The most common member of this group is the *Physalus*, so well known to all sailors under the name of the Portuguese Man-of-War. It is buoyed up by air bladders—in fact, its entire body appears as one bladder, which the animal is enabled to contract or expand at will. At first glance the *Physalus* appears to belong to quite a different family—suffering under some maltreatment; for from its lower side, what seem a number of entrails, of all shapes and sizes, hang down. When the upper surface or crest of its swimming bladder projects above the waves, it has a beautiful appearance, spangled with rays of purple, blue, and gold. This formation acts as a kind of sail, by means of which the creature is enabled to glide along with considerable speed.

This Physalus is a somewhat mysterious being, and zoologists have not as yet been able to determine many points connected with its structure and development.

The *Cirrigrade* Acalephæ, too, are a singular family. They exhibit a higher stage of development than those already alluded to, and possess a kind of skeleton embedded within their gelatinous bodies.

The *Porpita* and *Velella* are examples of this class, but for detailed descriptions I must refer the reader to larger works which treat on the subject.

I cannot conclude this brief and imperfect sketch of the Acalephæ without noticing their singular mode of reproduction. Nothing can appear more marvellous than this process when first brought before one's attention. It far excels the wildest dreams of fiction; and were it not so well authenticated by naturalists who have devoted labour and valuable time to gain ocular demonstration of the fact, we might well hesitate to believe the statements laid before us in their works.

For example, a Polype, as *Hydra Gelatinosa* or *Hydra Tuba* (found on buoys, oyster shells, &c., long submerged), will, it may be in a simple aquarium, produce a number of small objects which, on being examined through the microscope, are found to be, not young Polypes, but Jelly-fish! In process of time, the latter, by a wondrous law of nature, will produce in their turn, not Medusæ, but Polypes!

'Imagine,' says Mr. Lewes, 'a lily producing a butterfly, and the butterfly in turn producing a lily, and you would scarcely invent a marvel greater than this production of Medusæ was to its first discoverers. Nay, the marvel most go further still, the lily must first produce a whole bed of lilies like its own fair self before giving birth to the butterfly, and this butterfly must separate itself into a crowd of butter-flies, before giving birth to the lily.'

Let me now, by entering briefly into detail, endeavour to make the reader acquainted with the leading features of this mysterious subject, known as 'the alternation of generations.'

The adult Medusæ, then, gives birth to a number

of oval gemmæ or buds, appropriately so called by most writers, which appear like minute jelly bubbles, covered with numberless vibratile cilia. These organs, ten thousand times more delicate, we may imagine, than the eyelashes of some infant member of fairy land, are ever in constant motion. The currents produced thereby serve to propel the little animal to some stray pebble or stalk of sea-weed, situated at a respectful distance from its gelatinous relative. On some such object the young bud attaches itself, and proceeds to vegetate.

The body gradually lengthens, and becomes enlarged at its upper extremity; from this portion of the animal four arms appear surrounding a kind of mouth. The arms lengthen, and are soon joined by four others. These organs, as also the inner surface of the lips and of the stomach, are covered with cilia, and become highly sensitive. They are used in the same manner as the tentacula of the Actiniæ, namely, for the capture of food. There is this difference, be it observed, between the two animals, that while the infant Medusæ labours incessantly to gain its daily meals, the zoophyte remains still, and trusts to chance for every meal that it enjoys.

Fresh sets of arms continue to be developed successively upon the little jelly fish, until the whole amount in number to twenty-five or thirty. 'And the body, originally about the size of a grain of sand, becomes a line, or the twelfth part of an inch in length.'

Thus far there appears nothing particularly striking or improbable in the history of the Medusæ; the next stage, however, exhibits matter for our 'special wonder.'

The young Acaleph now throws off its animal existence, and sinks into a plant or compound polype.

The lower part of the body swells, and from thence, what may be termed a *stolen*, is thrown out. On the upper surface of the stolen one and even two buds are often formed. 'As the bud enlarges it becomes elongated, and bends itself downwards to reach the surface of the stone to which the elongated extremity adheres; after this the attached end is gradually separated from the body of the parent. When thus detached, a small opening presents itself at its upper end, its interior gradually becomes hollowed out, and cilia grow upon it, and tentacula begin to sprout around the mouth, exactly in the same manner as in the buds formed on the upper surface of the stolens.'

Thus, from a single bud numberless other buds are formed, each being endowed with equally prolific powers. If the parent be cut in half transversely, the cut will close in, attach itself to some object, and produce stolens and buds! If cut longitudinally, and the cut edges be allowed to touch each other, they will again adhere, and exhibit no trace of their ever having been divided. If the cut edges of each division be not kept apart they will approximate and adhere together, and thus two separate animals will be produced, each gifted with the power of throwing out stolens and buds with the same prodigality as if they had never been disunited!

How long this budding process of necessity continues we cannot tell. It may be only during the winter season. These creatures in their perfect condition are generally found crowding our seas during the summer months; most probably, therefore, as Sars and Steenstrup state, it is at the commencement of spring that they undergo the last portion of this 'transformation strange.'

Still, this cannot be taken as a general rule. Dr. Reid, who for a period of two years kept colonies of Medusæ, and assiduously watched the various stages of their development, found that the larvæ of one colony, which was obtained in September 1845, did not split transversely into young Medusæ in the spring of 1846, as he expected them to do, but continued to produce stolens and buds abundantly.

On the other hand, the larvæ of the other colonies, which this gentleman obtained in July, began to yield young Medusæ about the middle of March. This process takes place in the following manner: A 'bud' having arrived at maturity, it becomes 'cylindrical,' considerably elongated, and much diminished in diameter, its outer surface being marked with a series of transverse wrinkles.

These wrinkles, or rings, which frequently amount

to thirty or forty in number, are first formed at the top, and slowly extend downwards. Gradually as these furrows become deeper, the tentacula waste away, and upon the margin of the upper ring eight equi-distant rays are formed. The process continuing, in the space of a fortnight or so each groove or ring is in like manner furnished with rays. The Medusæ now present an appearance exactly resembling a series of cups piled up one within the other. Strange to state, each little cup becomes eventually endowed with life! As the uppermost segment is completely developed, it rests upon the slender lips of the one beneath. It then glides off from its old resting-place, and swims freely about in the water. Quickly it aspires to the rippling surface above, and by a series of graceful evolutions accomplishes its object. Once among the dancing waves and exposed to the rays of a cheering sun, our little Medusa assumes its complete form; and as a beautiful *Modeera formosa*, it may be destined at some time or other to be the prize of an ardent zoologist, who, I venture to assert, could not compliment it in more poetical language than Professor Forbes has already done. This delightful author, describing the little gem in question, says, 'It is gorgeous enough to be the diadem of sea fairies, and sufficiently graceful to be the night-cap of the tiniest and prettiest of mermaidens.' Or as an adult *Cyanea capillata*, our once insignificant jelly-bag may perhaps appear, and by an exercise of its urticating powers, send some unhappy swimmer smarting and trembling to his home.

While the Medusæ column proceeds to throw off from its uppermost part living segments of itself, its lower half, or stem, continues to grow, but does not become ringed, for as the budding process ceases, the last formed cup rests on newly-formed tentacula! Then again stolens are thrown out, on which young Medusæ are formed, as before described.

Contemplating such mysteries as these, the mind becomes bewildered and the spirits humbled.

> 'Imagination wastes its strength in vain,
> And fancy tries and turns within itself,
> Struck with the amazing depths of Deity.'

The above may be deemed one of the most interesting zoological theories that has ever been promulgated in modern times. It was founded by Chamisso, and termed the 'alternation of generation,' but was much improved and extended by the researches of Steenstrup. Professor Owen, however, had previously reduced the theory to a fixed and definite scientific form, under the title of '*Parthenogenesis.*' Another author, not viewing the Medusæ in the various stages of development as an aggregation of individuals, 'in the same sense that one of the higher animals is an individual,' proposes that each Medusa be considered as an individual, developed into so many 'zooids.'

Into this abstract question, of course, I cannot enter. The reader who would wish to know more of the subject than I have faintly shadowed forth in this chapter, may consult Steenstrup's Memoir, published by the Ray Society; Dr. Reid's admirable papers in the 'Magazine of Natural History' 2d series; Lewes' 'Sea-Side Studies;' and the learned works of Professor Owen.

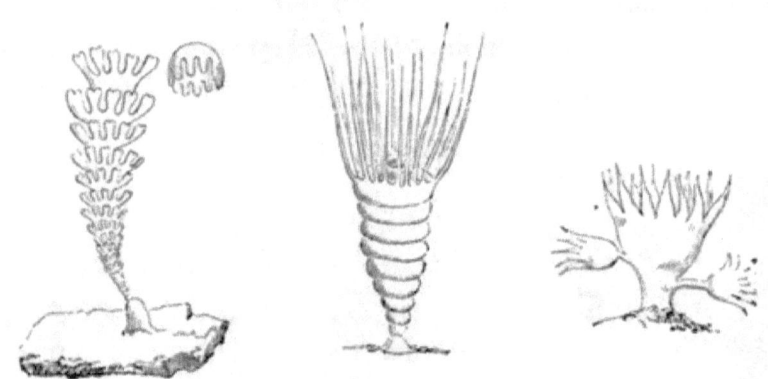

MEDUSÆ IN VARIOUS STAGES OF DEVELOPMENT.

CHAPTER XIV.

Doris, Eolis, &c.

'The inhabitants of the watery element were made for wise men to contemplate and fools to pass by without consideration.'—ISAAK WALTON.

XIV.

ONE fine morning during the month of January, on peering into my largest tank I perceived, attached to the upper portion of the marble arch (centre piece), a peculiar object that had evidently been deposited during the preceding night, but by whom or by what means I knew not. It resembled a fungoid growth, or riband of flesh, plaited up and attached at one edge to the stone. At every undulation of the water the object moved to and fro with an extremely graceful motion.

By careful and close examination it was seen to be covered with a film, that gradually expanded until it burst, and thus gave means of escape to thousands of minute white granules. On submitting these to the microscope, a most wondrous sight met my astonished eyes. Each dot or granule proved to be a transparent shell, resembling the periwinkle or rather the nautilus in shape; containing an animal whose excited and rapid movements were amusing to witness. From out the opening of the shell appeared now and again two rings of cilia. When these organs were about to be put into action, they reminded me of two circular tubes of gas connected together, and each containing innumerable perforations, which were sometimes suddenly and entirely lit up by a torch being applied to one end.

The *cilia* may be distinctly seen to play at a certain point, and then gradually extend round the circumference of the rings. When the whole are in full action their movements are so extremely swift *as to appear devoid of motion*, and thus bear a resemblance to rings of flame.

The result of the movements of the cilia was always evident in the vigorous evolutions of the little embryos, for the microscope filaments, while in action, caused the animal to roll about in all directions in a confined circle. When this envelope burst, the little nautiline dashed out, and then—then it was of little use attempting to get a view of the animal again, so rapid and violent were its movements to and fro, never resting for one instant on any spot, and least of all the spot wished. By the aid of blotting-paper, I sometimes reduced the quantity of water in the watch glass; and the animal, thus being compelled to confine its evolutions to a narrower stage, was more often within the field of view.

On visiting the sea-shore shortly after the discovery of the egg cluster just described, I perceived attached to numberless stones and large boulders thick clusters, composed of the self-same objects! Beside them were lying confused heaps of *sea-slugs*, evidently exhausted with their hatching exertions. Anything more repulsive to the eye than those animal heaps exhibited it would be difficult to conceive. Yet, at the same time, I know of no sight more pleasing than to watch the *Doris* in its healthy state, gliding along with outspread plume on the under surface of the water, or up the sides of the tank, more especially if it be observed through a powerful hand lens.

These remarks will perhaps convey some new information to the young naturalist, embracing as they do the leading facts connected with the wondrous embryotic development of many marine animals. The reader will already be prepared to learn that the vivacious little animal, moving by aid of cilia and enclosed in a *shell*, was in reality the youthful stage of that slow creeping gasteropod the *Doris*, which, in its mature form, is possessed of no cilia, nor any shelly covering whatever.

I should not have introduced this subject so familiarly did I not feel anxious to make my readers aware how easy it is for each of them to conduct experiments in the early stages of embryotic development, and to gain practical evidence of the wonders which this study unfolds.

'What,' eloquently asks Mr. Lewes, ' can be more interesting than to watch the beginnings of life, to trace the gradual evolution of an animal from a mass of cells, each stage in the evolution presenting not only its own characteristics, but those marks of affinity with other animals which make the whole world kin? To watch the formation of the blood-vessels, to see the heart first begin its tremulous pulsations, to note how life is from the first one incessant struggle and progress—these keep us with fascinated pertinacity at our studies.'

The remarkable fact above mentioned, of the young of the Nudibranchiate Gasteropoda being furnished with a shell is exhibited not only in Doris, but in Eolis, Tritonia, Aplysia, &c., while the embryos of the Purpura, Nerita, Trochus, &c., are likewise in their youthful state furnished with cilia, by the agency of which these animals swim freely about in their native element.

There is one exception to this, which occurs in *Chiton*, the early stage of which has recently been shown by the observations of Mr. Clarke and Professor Loven to be peculiar, and more resembling that of an annelid than of a mollusc. In this case the animal can scarcely be said to undergo a metamorphosis; for the embryo, even within the egg, has nearly the form of the parent, and the appearance of the shell-plates is a mere matter of development.

I have never attempted to count the ova that were contained in any single riband of Doris spawn, in fact I considered the task an impossibility; but at a rough calculation, I concluded there would not be less than a million. Dr. Darwin, however, travelling in the Falkland Isles, met with a riband of spawn from a white Doris (the animal itself was three and a half inches long), which measured twenty inches in length, and half an inch in breadth! and by counting how many balls were contained in a tenth of an inch in the row, and how many rows in an equal length of riband, this gentleman reckoned that upon a moderate computation there could not be less than six millions of eggs. Yet, in spite of such amazing fecundity, this Doris was not common. 'Although,' says Dr. Darwin, 'I was searching under the stones, I saw only seven individuals. No fallacy is more common with naturalists, than that the numbers of an individual species depend on its powers of propagation.'

This apparent paradox is not difficult of explanation when we consider the number of enemies which are always hovering near, and ready with hungry mouths to snap up the infant embryos as soon as they begin to show signs of vitality. The Hermit-Crabs are especially fond of Doris spawn, so much so, indeed, that the writer could never retain any for hatching purposes while any of the Paguri were near. Mr. Peach says they (the young Dorides) have myriads of enemies in the small *Infusoria*, which may be noticed, with a powerful microscope, hovering round them, and ready to devour them the instant weakness or injury prevents their keeping in motion the cilia, which serve both for locomotion and defence. Let them cease to move, a regular attack is made, and the animal is soon devoured; and it is interesting to observe several of the scavengers sporting with the empty shell, as if in derision of the havoc they have made.

The same difficulty of calculation does not exist, at least to any such extent, with the spawn of Eolis, which is laid in stringy coils. M. Gosse mentions a specimen of *E. papillosa* that laid nine strings of spawn in his tank between the 20th of March and the 24th of May, all as nearly as possible of the same length. Each string contained about a hundred convolutions, each convolution about two hundred ova, and each ovum including, on an average, two embryos, making a total progeny of forty thousand, produced from one parent in little more than two months.

I may mention that on no occasion have I ever found the spawn coils of either Doridiæ or Eolididæ in my tanks, or at the sea-shore, except during the months of January and February or March; neither have any of my specimens spawned more than once during an entire season. From noticing the same group of parent slugs congregated, and remaining, as I can affirm, for weeks near their egg clusters, evidently in a most enfeebled condition, it has occurred to me that on the Frith of Forth, at least, vast numbers of these animals do not long survive the hatching season.

Whether this be the case or not, it is a most singular fact that in this locality, a Doris more than one or two inches in length is scarcely ever to be met with.

There is at present in one of my tanks a specimen of the Doris of a pearly-white colour, a second, tinted white and pink, and two others which are quite *black*,—all being procured from the coast near Edinburgh. The last-mentioned animals are, I think, somewhat uncommon. When watching one of them in motion while the sun is shining down upon it, the hue of the creature changes from a black to a very deep purple, owing, no doubt, to its fleshy disc being many shades lighter than its body, which, being extended, and exhibited under a full glow of light, becomes semi-transparent. This peculiarity is not evident, of course, when the Doris is lying in a passive state, with all its gill-plumes closed up.

This sombre-coated gasteropod, although rare in some localities, is very plentiful in foreign parts, if the following may be received as an accurate narrative. ' On a reef of rocks near the island of Raiatea is a huge unshapely black or brown slug, here called '*Biche*,' from six to seven inches long, and five to six broad. Is is caught in vast quantities, and not only regarded as a great delicacy by the natives, but being cured, has become a valuable article of commerce in the China market, whither it is carried from many insular coasts of the Pacific by American ships. We have seen a number of lads fill three canoes in two hours with these sea-snails.'

Thus uninviting as this slimy animal seems to our English taste, there is evidently no doubt of its being used by the Chinese as an article of food, and according to the evidence of certain authors, is esteemed by the 'barbarians' a high-class luxury; but then we must remember that the inhabitants of the land of gongs and chopsticks, have always been famed for their singular gastronomic tastes. One poet writes:—

> 'That man had sure a palate covered o'er
> With brass or steel, that, on the rocky shore,
> First broke the oozy oyster's pearly coat,
> And risked the living morsel down his throat.'

But, 'Mandarins and Pigtails,' what was such *risk*, I ask, compared to that which *he* endured, who swallowed the first mouthful of birds'-nest soup? or horror of horrors, the first spoonful of sea-snail stew? Yet we are told that both the 'mucilage' and the *Bêches de Mer* dishes are savoury and highly grateful to the palate of an appreciating gourmand.

A recent author, describing a Chinese dinner from personal observation, tells us, that when the first dish, composed of birds'-nest soup, was over, he waited the advent of the next course with very nervous excitement. 'It was a stew of sea-slugs.

They are slippery, and very difficult to be handled by inexperienced chopsticks; but they are most pleasant and succulent food, not at all unlike in flavour to the green fat of the turtle. If a man cannot eat anything of a kind whereof he has not seen his father and grandfather eat before him, we must leave him to his oysters, and his periwinkles, and his craw-fish, and not expect him to swallow the much more comely sea-slug. But surely a Briton, who has eaten himself into a poisonous plethora upon mussels, has no right to hold up his hands and eyes at a Chinaman enjoying his honest, well-cooked stew of *Bêches de Mer*.

'During the discussion of this dish our Chinese master of the ceremonies solemnly interposed. We were neglecting the rudiments of politeness, no one had offered to intrude one of these sleek and savoury delicacies, deeply rolled in sauce, into the mouth of his neighbour. Efforts were made to retrieve the barbarian honour, but with no great success, for the slugs were *evasive*, and the proffered mouthful was not always welcome.'

CHAPTER XV.

The Crab and the Dainty Beggar.

'In taking a review of most, if not all the actions of the animal world, it must be obvious that, whether we allow them reason or not, the actions themselves comprehend those elements of reason, so to speak, which we commonly refer to rational beings, so that if the same actions had been done by our fellow-creatures, we should have ascribed them without hesitation to motives and feelings worthy of a rational nature.'—SCHLEIDEN.

'All things are bigge with jest; nothing that's plain
But may be wittie, if thou hast the vein.'
— GEORGE HERBERT.

XV.

I HAVE been observing for several days the movements of a Common Shore-Crab, which has been almost all his life under my protection. Although his present dimensions would render such a feat impossible, when first I shook the little fellow off a bunch of *C. officinalis*, he could have crawled with the greatest of ease into the mouth of a small pop-gun. We all know that members of this family are bold and daring in their attacks upon their weaker neighbours; upon each and all they wage a constant predatory warfare. The poor Pholades, however, are the favourite objects of their attacks. On these innocent bivalves the Crustacea successfully prey, unless they are protected by their usual rock-bound citadel, which, of course, they cannot always be. In order to watch the Pholas at work, it is necessary that the siphons should be more or less protruding from his tubular dwelling. If supported, say, to the full depth of his valves, the animal is secure; for I notice that neither crab nor fish can tear away the gristly ends of the siphonal appendages when withdrawn; and when disturbed, the poor Pholas leaves only this part in view. I have frequently seen the Fiddler-Crab embrace a Pholas in his claws, and struggle to pull him from his seat. On one occasion this operation was performed successfully, much to my annoyance, as I had been at some trouble to saw the rock away in order to watch easily every movement of the animal within. At night when I looked into the tank my pet was safe; next morning it was wedged under some pebbles, and the crab was feasting leisurely upon his tender flesh.

It is most amusing to watch the Blennies, too, attack a Pholas, cast into the tank, and to witness their mode of pecking at and shaking their victim, and turning innumerable somersaults with it in their mouths. The strength they exhibit in these manœuvres is perfectly astounding.

About two hours after they had received one of their favourite 'muttons' to feast upon, I peeped into the aquarium, and found, as I expected, the Blennies hard at their work of destruction. Behind them, among some bushy tufts of *I. edulis*, the little crab, before alluded to, was seated. In his arms he held an object unlike anything I had seen on sea or land. It appeared like a slender stick of beautifully iridescent opal. My amazement at this sight may readily be conceived, for I had not the remotest idea as to how he had become possessed of such a prize.

Next day I placed another devoted Pholas in the tank, and after a while looked in to see how its finny enemies were conducting themselves, when, what should meet my eye, but the crab, sneaking off with another opal baton in his arms! I was more puzzled than ever. It was quite certain that the object in question had been procured from the Pholas, yet I had not heard of, or ever seen anything like it in that animal.

I was 'on thorns' until next day, so that I might by watching solve the mystery. A third Pholas was flung into the den. The fishes, eager as usual, instantly attacked and pulled the mollusc to pieces. After a while the crab began to move about to and fro, evidently very restless, and anxious for my departure. I did retire, but only to such distance as would allow me a distinct view of his movements. In a few moments he stepped out mincingly on the tips of his toes, and crossed the tank to the spot where the poor Pholas lay, like some fine beau in Queen Anne's reign tripping jauntily down the Mall, or across St. James' Park, to feed the ducks in Rosamond's Pond.

The Blennies darted off at his approach. He then seated himself before the mangled corpse, and scraped at it vigorously, manifestly searching for some coveted treasure. Shortly after, perceiving him clutch at something, I quickly approached and disturbed his movements—took up the Pholas, and to my surprise found, on drawing out an object that protruded from the foot of the animal, that I possessed the pearly and gelatinous cylinder, such as the crab had twice before devoured with such evident relish.[1]

It was plain then that the little rascal had become so dainty, that he 'turned up his nose,' or rather his 'pair of noses,' at what is vulgarly termed the 'first cut,' and condescended only upon the tit-bits, for his marine banquet. So his crabship, in order to save himself trouble, actually waited until the fishes had cut up the Pholas to a certain point, when he would rush forward and seize on his favourite fare.

Some of my readers will doubtless remember the anecdote of the crossing sweeper, whose idiosyncrasy led him to covet diurnally a mutton-chop situated in the middle of the loin. My Lady Pepys, or Mr. Saccharine, the great grocer, couldn't always procure the desired 'cut!' *n'importe* the knight of the besom met with no such disappointment.

This individual's place of business was luckily situated opposite to a noted butcher's shop, which circumstance easily enabled him to watch until, from the demands of sundry customers, the perspective of the loin, which lay temptingly upon the chopping block, had become adapted to his point of sight. He would then step in and meekly order a simple pound avoirdupois. With this *bonne-bouche* care-

---

[1] My first introduction to the Hyaline stylet as above narrated, occurred in October 1858.

fully packed in his pocket, he would again mount guard, and remain until night. At dusk of evening he shut up shop,—that is, he swept the dirt over the parallel path that he had all day kept scrupulously clean, and then marched off to enjoy his dinner at a fashionable hour, in private.

Are not these cases palpably alike? Passing by certain details, were not the pawkiness and cunning of the epicurean beggar fully equalled by our diminutive friend, *C. mœnas*?

'But,' you ask, 'what then was the opal stick?' Ay, there's the rub, for even the greatest naturalists cannot positively agree as to the use and purpose of this mysterious organ. Yet it is to be found (as we have seen) in the foot of the Pholas, in the Mussel, the Cockle; and, in fact, it occurs in almost all bivalves both great and small. It is termed the 'hyaline cylindrical stylet,' and is very *lucidly* and scientifically described as 'an elastic spring to work the corneous plate or attritor, and by the muscular action of the foot and body, to divide and comminute the food, and especially the minute crustaceous and testaceous alimentary matters received into the stomachial cavity. It appears then that this appendage acts as *a gizzard*, and the bivalve mollusca are thus supplied with a masticatory apparatus very analogous to the gizzards of some of the gasteropoda.'

Now, the simple fact that I have stated above, of the hyaline stylet being found in the *foot*, and not in the stomach, at once proves that it cannot possibly act as a *gizzard* to the Pholas, or any other bivalve in which it is known to exist.

In the succeeding chapter I shall endeavour, from personal observation, to shed a slender ray of light upon the function of the stylet.

CHAPTER XVI.

The Pholas, &c.

(ROCK-BORERS.)

'He that of greatest works is finisher
Oft does them by the meanest minister.'

XVI.

At certain parts of the Scottish coast, the 'dykes,' or walls built near the road-side, are constructed entirely of rough-hewn pieces of hard sandstone rock, brought from the neighbouring shore. Sometimes a dyke will extend for two or three miles, without presenting an isolated fragment, in which the honey-comb-like perforations of certain species of the boring Mollusca are not more or less apparent.

A fragment of soft sandstone lies before me, measuring three and a half inches in length, and two inches in breadth, which, small though it be, contains no fewer than seventeen cylindrical tunnels. Each of these exhibits so wonderful a skill in construction, that human hands could not surpass it, though aided by 'all the means and appliances to boot,' of mechanical agency.

It is generally stated that the Pholas never intrudes itself into the apartment occupied by a neighbouring 'worker.' The Pholas, however, often intrudes on its neighbour; and such intrusion is manifested in the small piece of stone alluded to in no less than four instances. Here let me observe, that it is not always the larger mollusc that bores through the smaller one; it as frequently happens that the latter deserves the brand of wanton aggressor. Both cases are common enough, and, indeed, must of necessity occur, wherever at any time a colony of various sized Pholades are clustered together in a small portion of rock.

A fragment of rock riddled by the Pholas is a much more pleasing sight than can be witnessed at the sea shore in connection with that animal under usual circumstances. For this reason: When visiting the habitat of the boring bivalves, a host of small circular holes are sometimes seen; at other times the surface of the same portion of the beach appears comparatively sound, and it is only by striking a smart blow with a hammer upon the ground, that we render scores of orifices instantly observable in all directions, from each of which is thrown a small jet of water. This phenomenon is caused by the Pholades in alarm retracting their siphons, which had hitherto filled the entire extent of the tunnels. At such a locality, if a piece of rock be excavated, various specimens of these boring molluscs, shrunk to their smallest possible size, will be discovered at the base of the cavities, which are invariably of a conical form, tapered at the top, and gradually enlarging as they descend.

It must be evident, then, that neither the likeness of the animal, nor the formation of its singular dwelling-place, can be seen by the casual wanderer along the sea-shore.

It will also be apparent to the intelligent reader, that when once the Pholas is located in a certain spot, he becomes a tenant for life; for never by any chance whatever, can the poor miner leave his rocky habitation by his own unaided exertions, even were he so inclined. As he grows older and increases in size, nature teaches the animal to enlarge his habitation in a proportionate and suitable manner.

During the period of the boring process, the orifice becomes clogged above the shell with the *debris* of the rock, and this, if allowed to accumulate would speedily asphyxiate the animal. To get rid of such an unpleasant obstruction, the Pholas retracts, and closes the end of its siphons, then suddenly extends the 'double barrelled' tube to its full length, until it reaches the entrance of the tunnel. This movement often repeated, causes portions of the pulverized stone to be forced outwards at each operation.

It is interesting to watch the curious manner in which the end of the principal siphon is alternately closed and spread out when it reaches the water, like a man inspiring heavily after any unusual exertion; it is then made again to descend, and renew its task, as above described.

In extracting that portion of the *debris* which is

deposited at the *base* of the cavity, below the body of the industrious miner, a different plan is adopted. Wherever a Pholas is at its labours, there are always deposited within a circumference of several inches round the tunnel, myriads of short thin threads, which are squirted out from the smaller siphon.

The nodules on examination are found to be composed of pulverized rock, which is drawn in at the pedal opening, and ejected in the above manner, thereby effectually clearing the lower portion of the orifice. It was suggested to me that these thread-like objects were the fœcal matters of the Pholas, but this idea was soon dispelled by the assistance of the microscope; and, moreover, from the fact that the threads are never visible when the animal is in a quiescent state, but only when it is busily engaged in its mysterious task of boring.

I was for some time puzzled to find any aperture in connection with the club-like foot of the Pholas (*P. crispata*), although several of the bivalves were sacrificed to the cause of science. But what the microscope and scalpel in this instance failed to unfold, attentive watching of the animal in the aquarium made palpably apparent, in the following manner.

I had on one occasion captured about a dozen Pholades, some of which were embedded in the solid rock, others detached.

The first mentioned, I knew would be quite safe among the blennies and crabs, from the untempting and unedible nature of their siphonal tubes. Very different was the case with the defenceless, disentombed specimens. These were intended as food for their finny companions, who happened to be particularly fond of a change of diet. My 'pack' had subsisted for some time on Mussels, and on such excellent food, had become impudent, corpulent, and dainty. But overgorged epicures though they were, I knew that although everything else failed, a 'real live' Pholas placed before them would serve to speedily whet their appetites.

A splendid specimen of the siphoniferous bivalve was dropped into the tank, the base of which it had no sooner reached, than the fishes, with eager eyes and watering mouths, came hovering like a flock of vultures round the welcome meal thus unexpectedly placed before them.

One rascal, who seemed to be cock of the walk, came forward and made the first grip at the delicate fleshy foot, that in appearance was as white as a newly fallen snowflake. The pedal organ was, of course, instantly and forcibly withdrawn, so much so, indeed, as to be almost hidden from view, except at its extreme base. In this position it remained for several seconds. When the finny gourmand again boldly advanced to take a second mouthful, to my intense surprise he was, apparently, blown to a distance of several inches. I could scarcely credit the evidence of my senses. Another and another of the fishes were in their turn served in like manner as their leader. In a short time, however, the poor mollusc failed to repulse his enemies, and finally fell a passive victim to their gluttonous propensities.

Now comes the important question, 'How is the boring operation performed?' How can this simple animal, with its brittle shell, and soft fleshy body, manage to perforate the sandstone, or other hard substances, in which it lives?

For hundreds of years this query has been asked, and various are the replies which from time to time have been given. Singular to state, although specimens of the Pholas, and its allies the Saxicavae, are to be procured in abundance in many parts of the kingdom, the subject is not even yet positively settled.

There have been many theories advanced, some the result of fancy or guess-work; others, of practical study. All these have their supporters, but none have, by common consent, been adopted by physiologists as the true one.

Having for several years made this subject a study of personal observation, I believe I may venture to state, that I have succeeded in casting a feeble ray of light upon it; and, although the result of my labours may not be deemed conclusive, I may at least claim some credit for my endeavours to clear up a most difficult, though deeply interesting point in natural history.

The various theories promulgated on this knotty point are generally classed under five heads: 1st, That the animal secretes a chemical solvent—an acid —which dissolves the substance in which it bores. 2d, That the combined action of the secreted solvent, and rasping by the valves, effects the perforations. 3d, That the holes are made by rasping effected by silicious particles studding the substance of certain parts of the animal. 4th, That currents of water, set in action by the motions of vibratile cilia, are the agents. 5th, and lastly, That the boring mollusca perforate by means of the rotation of their shells, which serve as augurs.

Of all the above, the first which is quite a fancy theory, seems to meet with greatest favour among certain naturalists. But as it is rather puzzling to find a chemical solvent, which will act equally upon sandstone, clay, chalk, wax, and wood, this hypothesis can only be looked upon by practical men as ingenious, but incorrect. Even were it proved that the animal really possessed the power of secreting an acid sufficiently powerful, the question naturally arises, How can the shell escape being affected in like manner with the much harder substance in which it is situated?

The second theory, or the combined action of rasping and the secreted solvent, is, for obvious reasons, equally objectionable.

The third theory, which endeavours to account for the wearing away of the rock by means of silicious particles situated in the foot and other parts of the animal, has been for some time proved to be erroneous, from the fact, that the combined skill of some of our best anatomists and microscopists has failed to discover the slightest presence of any particles of silex in the Pholadidæ, although these are believed to exist in other families of the boring acephala.

The fourth theory, that of ciliary currents as an accessory agent in boring, is worthy of greater consideration, chiefly from the evidence we possess of the immense power which the incessant action of currents of water possess in wearing away hard substances.

We come now to what may be considered the most important of the theories above enumerated, viz., the mechanical action of the valves of the Pholas in rasping away the rock, &c. This hypothesis is one which most naturally suggests itself to the mind of any impartial person, on examining, for instance, the rasp-like exterior of the shell of *Pholas crispata*.[1] But as I shall endeavour to show, although the shell forms the principal, it does not by any means constitute the *sole* agent in completing the perforating process.

---

[1] Specimens of this species, I may mention, have always formed the subject of my experiments, and therefore are alone alluded to in the following remarks.

Mr. Clark, a clever naturalist, considers with Mr. Hancock that the powerfully armed ventral portion of the *mantle* of the closed boring acephala is fully adequate to rub down their habitations, and that the theories of mechanical boring, solvents, and ciliary currents, are so utterly worthless and incapable of producing the effects assigned to them, as not to be worth dwelling upon for one moment. Mr. Clark, therefore, comes to the conclusion that 'the foot is the true and sole terebrating agent in the Pholas.' This 'fact' he considers to be 'incontestably proved,' for the following reason, viz., because he had discovered specimens of this bivalve with the foot entirely obliterated,—which phenomenon, Mr. Clark states, is caused by the animal having arrived at its full growth, at which period the terebrating functions cease; and as 'nature never permanently retains what is superfluous,' the foot is supposed gradually to wither away, and finally disappear.

This, I suspect, is another 'fancy' theory. Although I have excavated hundreds of Pholades, some of giant-like proportions, it has never been my lot to witness the foot otherwise than in a healthy and fully developed condition.

Another writer, having no opportunity of viewing the living animal, does not consider it difficult to imagine the Pholas 'licking a hole' with its foot, from the fact that he (Mr. Sowerby) managed to make 'a sensible impression' upon a piece of kitchen hearthstone. 'I had,' he says, 'not patience to carry the experiment any further, but as far as it went, it left no doubt on my mind that, with the foot alone, and without any silicious particles, without a chemical solvent, and without using the rasping power of its shell, our little animal could easily execute his self-pronounced sentence of solitary confinement for life.'

Such an inconclusive statement as this would, I feel certain, never have been penned, had its author been so fortunate as to have had opportunity of watching a Pholas at work.

But, as Professor Owen truly observes, 'Direct observation of the boring bivalves in the act of perforation has been rarely enjoyed, and the instruments have consequently been guessed at, or judged of from the structure of the animal.' Such, evidently, is the case with Mr. Sowerby, and several other writers who treat on this subject.

Here we may call attention to the folly of naturalists endeavouring to tag a pet theory upon all the boring acephala, to the exclusion of every other. Such a system is defended upon the principle that, 'it is much more philosophical to allow that animals, so nearly allied as these in question, are more likely to effect a similar purpose by the same means, than that several should be adopted. Surely this is more consistent with the unity of the laws of nature, and that beautiful simplicity which is everywhere prevalent in her works.'

How much more shrewd and philosophical are the opinions of such a man as Professor Owen, who, when speaking of the mechanical action of the valves of *P. crispata,* says, 'To deny this use of the Pholas shell, because the shell of some other rock-boring bivalves is smooth, is another sign of a narrow mind.' Again, this learned author forcibly remarks, in direct opposition to the writer previously quoted, '*The diversity of the organization of the boring molluscs plainly speaks against any one single and uniform boring agent at all!*'

The more I study this subject, the more does the truth of the last-mentioned statement become apparent to my mind.

An examination of engravings of the shells, or even of the Pholas itself, when lying loose in the tank, or quietly seated in the rock, extending and retracting its siphons, fails to give one the slightest idea of its extraordinary appearance when enlarging its dwelling. At such times it seems to be a totally different animal, and to have suddenly acquired a most marvellous degree of power, energy, and perseverance, forming a striking contrast to its usual quiet, passive habits.

In the first place, as I have elsewhere written, it retracts its tube to, and even under, the level of its shell, just as a man, about to urge onwards some heavy mass with his shoulders, would depress his head to increase and concentrate his muscular power.

Then follows an expansion of the neck or upper part of the ventral border, from whence the siphons protrude. This movement closes the posterior portions of the valves below the hinge, and brings their serrated points together. The next act on the part of the animal is to place its foot firmly at the base of the hole; when leaning forward, it makes a sweeping movement fully half round the cavity, pressing firmly upon the umboes, which nature has strengthened for the purpose by two curved teeth fixed on the inside of the valves. At this stage it again reclines on its breast, and tilting up the shell as much as possible, it makes another motion round to its former position, leaning upon its back. By these intricate movements, which the Pholas appears to accomplish by a contraction almost painfully strong, it opens the rasping points of the valves. These execute a very peculiar scooping movement at the base of the cavity, and the animal having got so far, prepares itself for further exertion by a short rest.

The specimen whose movements I have attempted to describe, lived in my possession for a considerable time. It bored so completely through the piece of rock in which it was embedded, that the whole of its foot dropped through the aperture, and remained in this position for months, the animal, in consequence, being unable to change its position even in the slightest degree. Each movement of this specimen, both before and while the hole at **the base of the** cavity was gradually being **enlarged, was** watched, and every striking and interesting feature that occurred noted **down at the moment.** Various queries were put and answered, as far as possible, by direct ocular demonstration of the labours of the animal in the **vase** before me.

I consider myself to have been singularly fortunate in being able to view the actions of the creature from beneath, in consequence of the hole being bored through the rock. This circumstance allowed me distinctly to see what was going on **at the base of the** orifice.

My early observations have fortunately been confirmed in other captive Pholades, which at **various** periods have been domesticated **in my tanks.**

I am convinced, then, that the shell **forms the** *principal* agent in boring the animal's dwelling, without either acid or flinty particles. The late lamented Professor Forbes held that if this were the case, the rasping points on the surface of the valves would soon be worn down,—an appearance which, he says, is never seen. With all respect for such an eminent name, I must state that he was in error. Not only are the edges at certain times worn, but the rough surface is worn nearly smooth, appearing in certain parts of a white colour, instead of a light drab, as usual.

But the reader may ask, if certain parts of the valves are occasionally worn smooth, and the animal works so vigorously, how is it that they are never rasped through? This is a very natural question, and one that I put to myself repeatedly.

I have made frequent and careful observations while the animal was actually at work, in order to satisfy myself upon this point, and have always perceived that the particles of softened rock fell from, and on each side of, the large and well-developed *ligament* that binds the hinge, and extends to the lowest points of the valves. Moreover, this leathery substance always seemed scraped on the surface. I cannot, therefore, but believe that the ligament aids very materially in rubbing off the rock, or at all events, in graduating the pressure of the valves during the process, and that this curious organ, instead of being worn away, may, like the callosity upon a workman's hand, increase in toughness the more labour it is called upon to perform.[1]

The reason why so few specimens of the Pholades exhibit a worn shell may be thus explained: As the animal only bores the rock in sufficient degree to admit of its increased bulk of body, it only requires to bore occasionally, and there may be often an interval of many months, during which time nature

---

[1] Mr. Clark says, 'M. Deshayes, in his comment on Pholas, in the last edition of Lamarck, mentions the hinge as scarcely existing, and not being *a veritable ligament*.' How different from the fact; and I will observe, that '*if there is a genus better provided than any other of the bivalves with ligamental appendages, it is Pholas.* . . . . . *The Pholas is iron-bound as to ligament*, which in it is far more powerful in securing the valves, than is the shell of any other group of the acephala, of similar fragility and tenuity!'

may have renewed the serrated edge and rough surface of the valves, and thus enabled the creature to renew its wondrous operations.

We now come to a consideration of the foot, which, as many writers aver, forms the 'sole terebrating agent.'

Although this sweeping statement is incorrect, I will freely admit that the foot constitutes an agent second only in importance to the shell of the animal. A casual examination of any Pholas perforation will show that the foot could not have been the only instrument by which the cavity was formed, from the peculiar rings that line the lower portion of its interior. These rough appearances, I feel convinced, could be formed by no other means than the rotatory motion of the shelly valves.

The valves, however, could not rotate and press against the surface of the rock, were it not for the aid which the foot affords to the animal, by its being placed firmly at the base of the hole, and thus made to act as a powerful fulcrum.

This supposition fully accounts for the lowest extremity of the rocky chamber being always smooth, and hollowed out into a cup-like form by the action of the fleshy foot above alluded to.

The foot for a long time was a complete puzzle to me: I was unable to satisfy my mind as to how it acquired its seeming extraordinary power. The phenomenon was fully explained when I became aware of the presence of that mysterious organ the hyaline stylet, situated *in the centre of the foot*. The use of this springy muscle, which is, as we have shown in the previous chapter, by naturalists erroneously considered to be the gizzard of the animal, is, I believe, *solely to assist the Pholas in its boring operations.*

Perhaps some of my readers would like to know how to procure a sight of the stylet; if so, their wishes may be easily gratified. Take up a disentombed Pholas in your hand, and with a sharp lancet or point of a pen-knife, briskly cut a slit in the extreme end of the foot, and, if the operation be done skilfully, the object of your search will spring out of the incision to the extent, it may be, of a quarter of an inch. If not, a very slight examination will discover the opal gelatinous cylinder, which may be drawn out by means of a pair of forceps.

When extracted and held between the finger and thumb by its smaller end, the stylet will, if struck with a certain degree of force, vibrate rapidly to and fro for some seconds, in the same manner as a piece of steel or whalebone would be affected, under like circumstances.[1]

So long as a Pholas exhibits only the ends of its

---

[1] In the *Athenæum* (Nos. 1632 and 1636), were kindly published two letters from the author on the above subject, under the respective dates January 26th, and February 28th, 1859.

siphons to the eyes of a greedy crab, it is perfectly safe from attack. It is only when the fleshy foot is unprotected that it falls a prey to some hungry crustacean.

The toughness of the siphonal orifices is, I believe, a most important point, for, as I shall endeavour to explain, the siphonal tubes constitute important accessory excavating agents, to those already enumerated.

We all know that the hole which each young Pholas makes, when first he takes possession of his rocky home, is extremely minute,—not larger than a small pin's head; now, it stands to reason, that if the shell was the only terebrating agent, the opening of the cavity in question would always remain of the same size, or, perhaps, on account of the action of the water, a slight degree larger than its original dimensions. Such, however, is not the case.

Here is a fragment of rock exhibiting several Pholas holes. The aperture of one of these, which I measure, is nearly half-an-inch in diameter, while in juxtaposition with it is situated another cavity, measuring across the entrance less than the eighth part of an inch. The reader will at once perceive, if the foot and shell were the sole augurs, that as the animal descended deeper into the rock, the siphonal tube, as it enlarged in proportion to other parts of the animal, would have to be drawn out to an extremely fine point to fit the opening of the tunnel. But as this is not the state of matters, the conclusion forces itself upon us, that that portion of the orifice situated above the shell of the animal must be enlarged by the constant extension and retraction of the siphons, aided by currents of water acting on the interior surface of the cavity.

This same theory will also serve to explain how it is that all Pholades situated at the same depth in the rock, are not all of a uniform size. I have frequently seen a piece of rock exhibit the peculiarity of two burrows of vastly different proportions as regards breadth, being precisely the same depth from the surface of the stone. This appears to me equally wonderful and puzzling at first sight, as the ' boring' question.

What age is attained by any species of the rock-borers before they arrive at full growth, there are no means of knowing. This point, like several others in the history of these animals, still remains a mystery, nor is it likely soon to be cleared up. The largest specimen of *P. crispata* that I have seen is at present in my possession. Each valve measures three and a half inches in length, by two inches in breadth. Some foreign specimens of this species, and especially of *P. dactylus*, are, however, frequently found of much larger dimensions.

On no occasion have I ever examined any Pholas excavation that had lost its conical shape, a fact that seems to prove that the successive stages of the boring operation must have taken place solely in consequence of the animal not having reached its adult form.[1] For had the shell attained its full development, and its owner continued to labour, and rasp away the rock, the sides of the cavity at its base would necessarily present a parallel appearance—a phenomenon which is never witnessed.

From this we may conclude that the depth of the perforation, which is seldom many inches, depends entirely upon the growth of the mollusc.

When keeping specimens of the Pholas for observation, the usual plan is to chip away the rock to the level of the valves, so that the whole of the animal's siphonal tubes may be distinctly seen, however slightly these organs may be extended. This plan, I found, did very well for a time, but I was annoyed to witness, that in the course of a few months, the siphons ceased to be either advanced or retracted, —they having become, as it were, rudimentary.

To obviate such contingency, the writer adopted the following scheme.

To place in the tank a Pholas completely embedded in a fragment of rock, so that nothing but the tips of its siphons, when extended to the utmost, were visible, would not afford much pleasure to the stu-

---

[1] The above remark holds good, even although (*as is frequently the case*) the animal wilfully deviates from the straight path, and bores its tunnel in a curved form.

dent. I therefore managed to saw away the rock in such a manner, as to leave a narrow slit along the entire length of the tunnel, so as to expose the slightest movement of the animal within. Having natural support for its siphons, I expected that these organs would be constantly retracted and extended; but such was not the case; at least for so long a period as I had anticipated.

After repeated experiments, I have now discovered that whether the siphons be protected as above described or not, they will always be vigorously exercised if the animal be placed in shallow water, so that its tubes when fully extended will reach the surface of the fluid.

The conclusion, from what has been stated, is, that the Pholas can no longer be considered a weak and helpless animal. Possessed of a rasp-like shell, a horny ligament, retractile tubes, a strong muscular foot, and a powerful spring or stylet, it is not by any means difficult to conceive that these agents when they are all brought into play, are fully equal to the task of excavating the rocky chamber in which the animal lives.

CHAPTER XVII.

The Sea-Mouse.

(APHRODITE ACULEATA.)

'For seas have . . . . .
As well as earth, vines, roses, nettles, melons,
Mushrooms, pinks, gilliflowers, and many millions
Of other plants, more rare, more strange than these,
As very fishes living in the seas.'

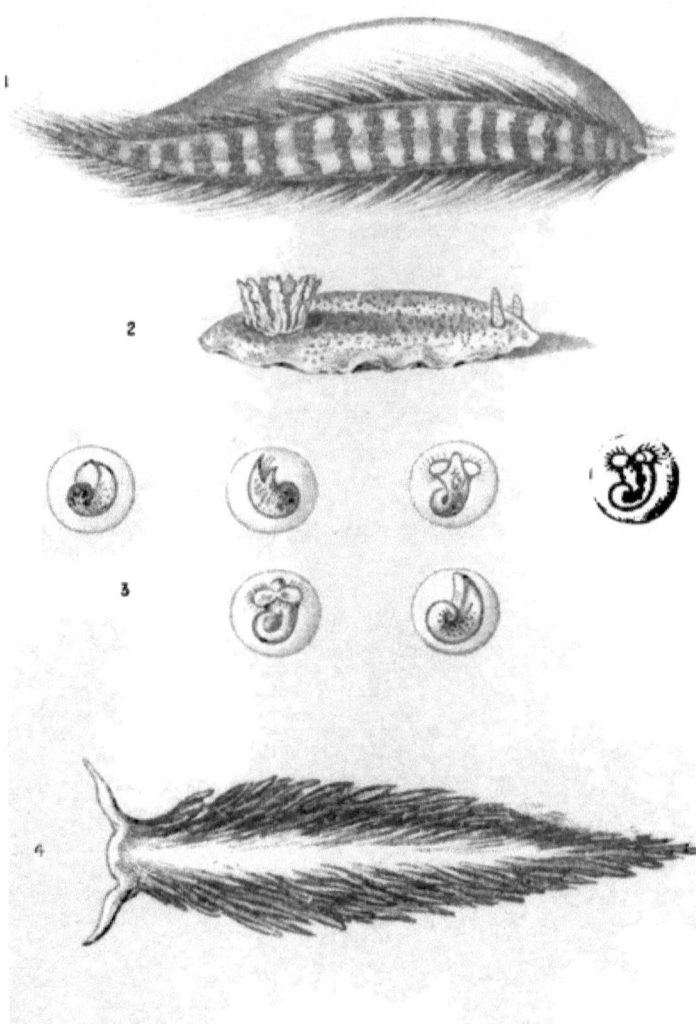

1 THE SEA MOUSE (Aphrodite aculeata)
2 THE DORIS
3 YOUNG OF THE DORIS
4 EOLIS PAPILLOSA

XVII.

BEAUTEOUS stars also the sea contains, as numberless, though not so brilliant in appearance as those which stud the firmament of heaven; flowers, too, grow beneath the wave, and rival in loveliness the gems which adorn our fields and hedge-rows. Nay, more, like the land, the ocean owns its various grasses, its lemons, and cucumbers, its worms, slugs, and shelly snails, its hedgehogs, its birds, its ducks and geese (*anatidæ*), its dogs, its hares, and lastly its *mice* (*aphroditæ*.) The latter objects, despite their unprepossessing name, being in no wise less interesting than those above mentioned.

The *Aphrodite aculeata* is, perhaps, one of the most gorgeous creatures that inhabits the seas of our British coast. Its body is covered with a coating of short brown hairs, but as these approach the sides of the animal, they become intermixed with long dark bristles, the whole of which are of an iridescent character. In one respect this creature bears no resemblance to its namesake of the land, being extremely slow and sluggish in its movements (at least according to our experience) when kept in confinement. Some writers, however, affirm that the Aphrodite possesses the power, although seldom exercised, of both running and swimming through the water with considerable speed.

In general the animal loves to tenant the slimy mud, and wherever the writer has happened to come upon a specimen at the sea-shore, its back has always been thickly coated with sand or dirt. The Sea-Mouse, then, unlike the peacock, can never be deemed an emblem of haughty pride, yet has nature in her lavish beauty endowed this humble inhabitant of the deep with a richness of plumage, so to speak, fully equal in its metallic brilliancy to that which decorates the tail of the strutting bird we have mentioned. As the bristles of the Aphrodite are moved about, tints—green, yellow, and orange, blue, purple, and scarlet—all the hues of Iris play upon them with the changing light, and shine with a metallic effulgence. Even if the animal, when dead, is placed in clear water, the same varied effect is seen as often as the observer changes his position.

Not only are the *Setæ* worthy of notice on account of their lustrous beauty, but also for their shape, and the important part they play in the economy of the animal. These lance-like spines seem to be used by the Aphrodite as weapons of defence, like the spines of the hedgehog or porcupine. In some species they are like harpoons, each being supplied with a double series of strong barbs.

The instruments can all be withdrawn into the body of the animal at will, but we can easily conceive that such formidable weapons being retracted into its flesh would not add to the creature's comfort—in fact they would produce a deadly effect, were it not for the following simple and beautiful contrivance.

Each spine is furnished with a double sheath composed of two blades, between which it is lodged; these sheaths closing upon the sharp points of the spear when the latter is drawn inwards, effectually guard the surrounding flesh from injury.

The shape of this animal is oval, the back convex, while the under part presents a flat and curious ribbed-like appearance. Its length varies from three to five inches; specimens, however, are sometimes to be procured, even on our own shores, of much larger dimensions.

CHAPTER XVIII.

Star-Fishes.

(OPHIURIDAE AND ASTERIADAE.)

'As there are stars in the sky, so there are stars in the sea.—LINK.

XVIII.

There are not a few persons still to be met with, who believe that man and the lower animals appeared simultaneously upon the face of the earth. Geology most forcibly proves the error of such an idea, for although the fossilized remains of every other class of organized beings have been discovered, human bones have nowhere been found. This fact, though deeply interesting, is perhaps not more so than many others which this wonderful science has unfolded. What can be more startling to the student for instance, than the information that for a long period, it may be thousands of years, no species of fish whatever inhabited the primeval seas? True it is that certain creatures occupied the shallows and depths of ocean, but these were of the lowest type. The most conspicuous were the coral polypes, which even then as now were ever industriously building up lasting monuments of their existence, as the Trilobites, a group of Crustacea, and the Crinoids, or Lily-stars.

The last-mentioned group of animals were analogous to the present tribe of Star-fishes, and are now nearly extinct. The body of the Lily-star, which resembled some beautiful radiate flower, was affixed to a long, slender stalk, composed of a series of solid plates superposed upon one another, bound together by a fleshy coat, and made to undulate to and fro in any direction at the will of the animal. The stalk was firmly attached to some foreign substance, and consequently the Crinoid Star-fish, unlike its modern representative, could not rove about in search of prey, but only capture such objects as came within reach of its widely expanded arms. 'Scarcely a dozen kinds of these beautiful creatures,' observes Professor Forbes, ' now live in the seas of our globe, and individuals of these kinds are comparatively rarely to be met with; formerly they were among the most numerous of the ocean's inhabitants,—so numerous that the remains of their skeletons constitute great tracts of the dry land as it now appears. For miles and miles we may walk over the stony fragments of the Crinoidae, fragments which were once built up in animated forms, encased in living flesh, and obeying the will of creatures among the loveliest of the inhabitants of the ocean. Even in their present disjointed and petrified state, they excite the admiration not only of the naturalist, but of the common gazer; and the name of stone lily, popularly applied to them, indicates a popular appreciation of their beauty.' Each wheel-like joint of the fossil Encrinite being generally perforated in the centre, facility is thus afforded for stringing a number of these objects together like beads, and in this form the monks of old, according to tradition, used the broken fragments of the lily-stars as rosaries. Hence the common appellation of St. Cuthbert's Beads, to which Sir Walter Scott alludes,—

> 'On a rock by Lindisfarn
> St. Cuthbert sits, and toils to frame
> The sea-born beads that bear his name.'

One solitary species of the Crinoid Star-fishes has of late years been found to flourish in our own seas; it is, however, affixed to a stalk (pedunculated) only in the early periods of its existence.

When first discovered by Mr. Thompson in its infant state, the *Pentacrinus Europæus* was believed to be a distinct animal. It was taken attached to the stems of zoophytes of different orders, and measured about three-fourths of an inch in height. In form it resembled a minute comatula mounted on the stalk of a Pentacrinus. Subsequent research has proved that the little stranger was merely the young state of the feather star *Comatula rosacea*, and that although for a certain period attached to a slender waving stem, the Pentacrinus, when arrived at a certain stage of development, feels fully able to start life on its own accord, and hence takes opportunity to break off its early ties, and become a free animal, dependent upon its own exertions for subsistence.

It is no uncommon thing, as a late writer forcibly remarks, in the inferior classes of the animal kingdom, to find animals permanently attached from the period of their birth, and during all their existence. Familiar examples of this occur in the oyster, and various other bivalve shell-fish, as well as in numerous compound zoophytes. We likewise meet with races which are free and locomotive in their first stages, and afterwards become permanently fixed; but an animal growing for a period in the similitude of a flower on a stem, and then dropping from its pedicle, and becoming during the remainder of its life free and peripatetic, is not only new, but without any parellel in the whole range of the organized creation.

The Comatula, or as it is commonly called, the Rosy Feather-star, is allowed to be without exception the most lively of all the star-fishes. Its movements in swimming are said to resemble exactly the alternating strokes given by the medusa to the liquid element, and have the same effect, causing the animal to raise itself from the bottom, and to advance back foremost even more rapidly than the medusa. It has ten very slender rays with numbers of long beards on the sides. The body, which is of a deep rose colour, is small and surrounded with ten little filiform rays. The extremities of these organs are shaped like claws, by means of which the animal attaches itself to various kinds of sea-weed, and other submarine objects.

The adult Comatula generally measures about five inches across its fully expanded rays.

Before treating of what are termed the *true* Star-fishes, we require to dwell briefly upon an intermediate family named by Professor Forbes the *Ophiuridæ*, 'from the long serpent or worm-like arms, which are appended to their round, depressed, urchin like bodies. . . . They hold the same relation to the Crinoidea that the true Star-fishes hold to the Sea-Urchins. They are spinigrade animals, and have no true suckers by which to walk, their progression being effected (and with great facility) by means of five long flexible-jointed processes placed at regular distances around their body, and furnished with spines on the sides and membraneous tentacula. These processes are very different from the arms of the true Star-fishes, which are lobes of the animal's body, whereas the arms of the Ophiuridæ are superadded to the body, and there is no excavation in them for any longation of the digestive organs.'*

The British Ophiuridæ are now classed under two genera; of the Ophiuræ, or Sand-stars only two species (*O. texturata* and *O. albida*) are found on our shores; and the Ophicomæ, or Brittle-stars, of which there are ten.

---

* British Star-fishes.

An extraordinary feature, characteristic of all the above-mentioned animals, is the great tendency which they have to mutilate themselves, and throw their limbs about in fragments on the slightest provocation. If a specimen be handled, a certain number of fragments will assuredly be cast off. If the rays become entangled in sea-weed, or even if the water in which the animal resides happens to become impure, the same disastrous result follows, until nothing but the little circular disc remains. As a set off against this weakness, both the Ophiuræ and the Brittle-stars possess reproductive powers of a high order. Hence it not unfrequently happens that if each and all the rays of a specimen be rejected, the animal will live on, and eventually, perhaps, become a complete and perfect star-fish.

The best means of preserving an Ophiura is to let the devoted animal remain for a time expanded in sea-water, then with a small pair of forceps lift it carefully up, and plump it into a bath of cold 'fresh' water, letting it lie there for about an hour. The animal speedily dies, as if poisoned, in the fresh liquid, in a state of rigid expansion. Some writers recommend that, at this stage, the specimen should be dipped for a moment into boiling water, and then dried in a current of air; but I have never been able to detect any great benefit arising from the adoption of the process.

When examining any of the Brittle-stars, I have

always found it an excellent plan to raise them up by aid of the forceps applied to the disc. By this means a specimen may be moved about without any fear of mutilation; whereas if the fingers be used as forceps, an unhappy result will assuredly follow.

The *Ophiocoma rosula*, figured on Plate IX., will serve to convey to the reader a general idea of this class of animals. Its popular title is the Common Brittle-star, indicative of the inherent fragility of the species, as also of their prevalent appearance at the sea-shore; but, though so exceedingly 'common,' we must at the same time in justice add, that the *O. rosula* exceeds in beauty many other species which are rare, and consequently more highly prized by the collector.

It is very abundant on all parts of the British coast, and is often found in clusters upon the stems of *L. digitata*, and as frequently upon the under side of boulders. In dredging, the Brittle-star is an unfailing prize. It is a marvellous sight when the scrapings of the ocean bed are spread out upon the dredging-board for examination, to see hundreds of these singularly delicate creatures twisting and twining about in all directions,—over each other's bodies, through the weed, sand, shells, and mud, and strewing fragments of their snake-like arms upon every surrounding object.

At the mere mention of 'Star-fishes,' the most uninitiated reader will at once realize in his mind's eye a tolerably correct notion of the form of these curious productions of the marine animal kingdom, even although he had never seen a living or dead specimen.

The body of the animal is divided into rays, like the pictured form of one of the heavenly stars, and the fancied resemblance is most apparent in the Asteridæ, or true Star-fishes, of which we are now about to speak.

This wonderful race of animals, for their beauty of colour, elegance of shape, and peculiarity of structure, possess a great degree of interest, not only to the naturalist, but also to the casual observer by the sea-side.

There are fourteen British species of Asteriadæ, which are arranged under four families, namely— the Urasteriæ, the Solasteriæ, the Gonasteriæ, and the Asteriæ. This group contains no less than eight generic types, clearly distinguished from each other by certain characters, 'derived from the outline of the body, the number of rows of suckers in the avenues, and the structure and arrangement of the spines covering the surface and bordering the avenues.'

There are four species of Star-fishes belonging to the genus Uraster, the most common of which is the *Uraster rubens*, or Common Cross-fish.

No person in the habit of visiting the sea-shore can be unfamiliar with the likeness of this creature, which is generally seen lying wedged in some rocky crevice, or among the Fuci, there patiently waiting the return of the tide.

At such a time, the Devil's-hand (as the Irish people term it), does not appear by any means attractive. If placed in water, however, its appearance becomes wonderfully improved.

Here is a small specimen, just brought from the sea-shore at Cockburnspath (a most romantic and delightful locality, situated on the coast of Berwickshire). It is neatly wrapped up in a mantle of seaweed. Freed of its verdant envelope, I deposit the youthful Rubens upon his back—'willy-nilly'—in a tumbler partly filled with clear sea water, and then proceed to watch its movements through a magnifier.

At a glance we perceive that each of its five rays is grooved on its lower surface, and filled with minute perforations, through which is gradually protruded a multitude of fleshy suckers, knobbed at the end. It is by aid of these organs that the animal grasps its food, and changes its position, as we shall presently see. One of the rays is now slowly lifted up and moved about in various directions, while from its extreme point the suckers are extended to the utmost limit. No sooner do they touch the side of the vessel than they are firmly fixed and contracted. A *point d'appui* being thus gained, the animal is enabled by degrees to draw its body round, so as to get another regiment of suckers into play, and, by such plan of operations being repeated, the animal is eventually enabled to 'right itself,' and crawl up the polished surface of the glass.

Generally, when the Star-fish is disturbed, or placed on a dry piece of stone, the suckers are withdrawn into the body, leaving no signs of their previous existence except a series of minute tubercles. In fact, the Asterias, although enabled to adhere with great tenacity to any foreign object when immersed in water, possesses but little power to retain its hold if the fluid be removed. Hence the young zoologist, keeping this peculiarity in mind, should not too hurriedly return a verdict of 'Found dead,' when he meets with a helpless specimen upon the beach, for in all likelihood, were the creature to be laid for a few minutes in a rock-pool, it would soon exhibit signs of returning animation.

A simpler, though not so sure a test for ascertaining whether a Star-fish be living or not, is to handle the specimen. If it feel soft and flabby, it is dead; but if tolerably firm to the touch, it may be 'recalled to life,' by the means pointed out.

It may not be out of place to chronicle here a singular circumstance which the writer has often verified in connection with the true Star-fishes. It is this. When any captured specimens have been placed in confinement, no matter how large or small such might be, they never moved through the liquid element with a tithe of the rapidity that I well knew they were capable of. At the sea-side, I have seen a specimen of the Cross-fish glide through the water so nimbly, yet withal so gracefully, that I have felt inclined to rank natation among the few other acomplishments of which the species can boast.

The *Uraster rubens* is also popularly known as 'Five Fingers.' For ages past it has been subject to the bitter denunciation of fishermen and others, for the injury which it is said to inflict upon oysters. At one time, according to Bishop Spratt, the Admiralty Court laid penalties upon those engaged in the oyster-fishing who did not tread under their feet, or throw upon the shore, a fish they call a Five-Finger, resembling a spur-rowel, because that fish gets into the oysters when they gape, and sucks them out. Poets have also endeavoured to perpetuate the vulgar opinion:—

> 'The prickly Star-fish creeps with fell deceit,
> To force the Oyster from his close retreat,
> Whose gaping lids their widened void display;
> The watchful Star thrusts in a pointed ray—
> Of all its treasures robs the rifled case,
> And empty shells the sandy hillock grace.'

Even yet the oyster fishermen at certain localities wreak all possible vengeance upon the 'submarine Dando's,' for their supposed gourmandizing propensities. I say *supposed*, for although so many naturalists have studied the question, it is not, up to the present time, satisfactorily settled. Some deny the alleged tendency altogether, while less sceptical observers are unable to understand the mode in which the Star-fish could injure an animal apparently so capable of self-defence as the oyster. According to certain authors, the Star-fish encircles the oyster with its five fingers, and by some clever process of suction destroys the unfortunate mollusc. Others, again, maintain that the first step of the attack is the injection of some marine chloroform between the shells of the oyster, and that during the insensibility that follows, the Star-fish effects an entrance.

As this is an interesting subject, perhaps the reader would like to have the exact words which are used by two celebrated naturalists, one of whom attempts to vindicate the character of the Asteridæ, the other to blacken it.

Sir John Dalyell—a high authority upon all matters of marine zoology—shrewdly remarks: ' I have not heard it suggested that the Star-fish possesses any kind of solvent compelling the bivalves to sunder. Neither can its hostility be very deadly to the larger univalves, from the distance to which they are enabled to retreat within their portable dwellings. Their general habits are, to force the shells of smaller bivalves asunder, and to devour the contents; they likewise consume the substance of ordinary fishes entire; nevertheless, as far as I am yet aware, their destruction of oysters is destitute of evidence. The Star-fish sometimes shows an ever-sion of stomach, or of some membrane of it. Whether this may be the means of affecting their prey, merits investigation.'

Professor Jones, who affirms that in the latter suggestion Sir J. Dalyell has nearly hit upon the true solution of the problem, thus gives what *he* considers to be the correct mode of procedure on the part of the Star-fish : ' Grasping its shell-clad prey between its rays, and firmly fixing it by means of its prehensile suckers, it proceeds deliberately to turn its stomach inside out, embracing in its ample folds the helpless bivalve, and perhaps at the same time instilling some torpifying fluid, for the shells of the poor victim seized soon open, and it then becomes an easy prey.'

Now, many fishermen with whom I have conversed hold the same opinion as Bishop Spratt, and believe that when the oyster is gaping the Star-fish insinuates a finger, and hastily scrapes out the delicious mouthful ; nay, further maintain that the Star-fish is far from being successful at all times, very often, especially when there has only been one ray inserted, the frightened oyster grasps it with all his might, and obliges his discomfited opponent to retire minus a limb.

If the writer might venture to suggest an opinion, he would express his belief that the following is the correct account of the state of matters. He believes with the fishermen that frequently the star-fish begins his attack by inserting an arm, but he does not believe that the oyster under such circumstances escapes with life. Let us suppose the star-fish to have succeeded in insidiously introducing a ray within the shell of the apathetic oyster, and that the oyster immediately resented such intrusion by closing his shell with all the force he can exert. The opposite argument at this stage is, that the intruder is obliged from *pain* to abandon his hold, and even pay for his audacity by the forfeit of a limb. But against this we advance the notorious fact, that the star-fish, like so many marine creatures of a similar organization, is remarkably indifferent to pain. I therefore believe the true explanation to be, that the oyster being unable to sustain such continued muscular exertion for nearly so long a time as the star-fish can tolerate the pressure upon its ray, the latter is consequently, in the long run, successful.

The number of rays in the several genera of the true Star-fishes is extremely various. In the genus *Uraster*, as we have seen, five is the predominant number. If we turn to the two species which comprise the genus *Cribella*, we still find the quintuple arrangement adhered to. In *Solaster endeca*, on the contrary, the rays vary from nine to eleven, and even reach as high as twelve or fifteen in *Solaster papposa*.

In the genus *Palmipes* we have the pentagonal form, it is true, but the space between each ray is filled up, so as to resemble the webbed foot of a bird, hence the popular title of this solitary species, 'The Bird's-foot Sea-star.' 'It is the flattest of all its class, and when alive it is flexible like a piece of leather.' Passing by the 'Cushion-stars' (which have five *angles*—it seems a misnomer to call them rays), which connect the true Star-fishes with the Sea-Urchins, we come lastly to the 'Lingthorn,' *Luidia fragillisima*, with its seven rays. This is the animal of which Professor Forbes discourses so pleasantly about its winking derisively at his despairing endeavours to preserve even a small portion of what at that time was his maiden specimen. The Luidia is even more brittle—more regardless of its wholeness, than the *Ophiuræ*, which renders the capture of a perfect specimen a most difficult task.

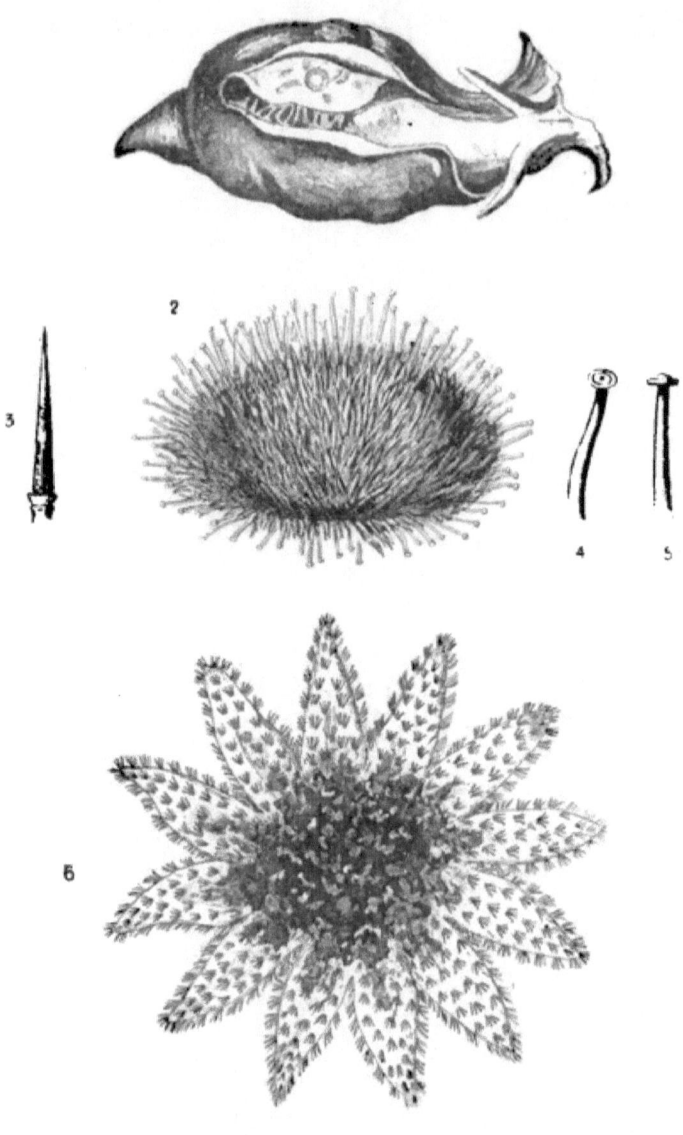

1 THE APLYSIA, or SEA-HARE.
2 PURPLE-TIPPED SEA-URCHIN.
3 Spine of D?
4,5 Suckers of D?
6 COMMON SUN-STAR

XIX.

SEA-URCHINS are frequently taken in dredging. Several common species, usually of a small size, are often found among the rocks situated between tide marks.

Into the aquarium no specimens larger than from one to two inches in diameter should be introduced, and even these require to be closely watched, for if afflicted with a fatal illness, I know of no animal whose remains sooner taint the water. Almost before life is extinct, the Urchin throws out a light-coloured nauseous fluid, that speedily poisons the surrounding water, and, of course, causes the destruction of any inhabitants of the tank who may neither have the sense nor opportunity to inhale copious draughts of fresh air. As a rule, if the suckers are motionless, or if on touching the animal it is found not to be adherent to any object, transfer it at once to your 'infirmary' for further observation.

I have always found small specimens to be much

more lively and walkative, (if I may be allowed the expression) than their more corpulent brethren.

The inflexible, mail-like crust, or shell, as it is commonly called, of the Echinus is perhaps one of the most marvellous objects on which the eye can rest. Although at first sight it appears to be a solid calcareous box, it is in reality composed of several hundred pentagonal plates,[1] of various sizes, so closely dove-tailed together that their marks of junction are scarcely perceptible. Upon a superficial examination we are apt (most erroneously) to consider this wonderful piece of work to be more elaborate than the wants of the animal demand. The fact of the Lobster or Crab throwing off its entire shell at certain seasons, to admit of the increased growth of the animal is a truly marvellous phenomenon, still, it would more excite our wonder were we to find that, instead of being cast away at all, the hard, inelastic envelope which surrounds the bodies of crustaceans was made to swell or expand proportionately with the soft parts of the animal! Now, the mosaic-like shell of the Sea-Urchin, though built up, as before stated, of several hundred pieces, is by a beautiful process slowly and imperceptibly enlarged correspondingly with the growth of the animal.

The gradual enlargement of the Echinus shell takes place in the following manner:—

---

[1] In a specimen that I examined, and then carefully took to pieces, there were exactly 1780 plates.

Over the entire surface of the globular shell, spines, and joints of the living Urchin, there exists a delicate membrane that insinuates itself between the pentagonal plates above mentioned, and continually deposits around the edges a certain portion of calcareous matter (carbonate of lime). The same process being also carried on by the fleshy covering that surrounds the spines, &c., it must be evident that so long as the vital power of the animal exists, each plate and spine, still keeping to its original form, must be daily and hourly augmented in size until the Sea-Egg has attained its full and mature dimensions.

As to how the spines retain their relative position in each plate, as the latter gradually becomes enlarged, I cannot positively state; but may be permitted to mention, that, judging from carefully prepared sections of the plates when submitted to the microscope, each spine appeared to my eye to be by some singular process urged along in a kind of groove to its proper place.

The hedgehog-like spines that surround the globose body of the Sea-Urchin are all moveable at the will of the animal,—each prickle being connected by a ball-and-socket joint to a pearly tubercle, which acts as the 'socket' on which the 'ball' of the spine revolves. If the spine be removed, a comparatively smooth surface will be left, on which are various sized tubercles systematically arranged.

Situated at regular intervals between the tubercles are ten broad bands, disposed in pairs, and containing many hundreds of very minute perforations, or ambulacral orifices, as they are generally termed by naturalists.

Through these apertures issue numerous suckerlike feet, closely resembling those of the Star-fish, but endowed with far greater powers of contraction and extension.

The number of suckers is very great. In an Urchin measuring exactly three inches in diameter, by aid of a hand lens, I counted no less than 3300 pores in the ten avenues. Now, these pores are always situated in pairs, and as each sucker occupies a pair of pores, it will give 1650 as the total amount of suckers.

There is no doubt that it is almost entirely by means of these curious organs that the Sea-Urchin is enabled to move about from place to place, although no less an authority than Professor Agassiz asserts to the contrary. 'How, in fact,' says this author, 'could these small tentacula, situated as they generally are in that part of the body which is never brought into contact with the ground when the animal moves, and overhung by calcareous solid spines—how, I ask, could these flexible tubes be used as organs of motion? It is an undeniable fact, and I have often observed it myself, that *it is with their spines the Echini move themselves, seize their prey, and bring it to their mouths* by turning the rays of their lower edge in different directions. But the correction of an error respecting the functions of the ambulacral tubes does not solve the problem relating to their nature and use. This problem we are yet unable to solve, as we know nothing more respecting them than that they are connected with the aquiferous system.'

Many other writers, among whom is Professor Forbes (from whose work on Star-fishes I have transferred the foregoing extract), assert, in opposition to the great Swiss naturalist, that the Echinidæ move by the joint action of their suckers and spines. 'The argument,' says the great British naturalist, 'against the suckers being organs of motion, founded on their position above as well as below, would equally apply to the spines, to which organs Professor Agassiz has attributed all progressive powers in these animals.'

The fact is now so well established, that it is scarcely necessary for the writer to state, that from personal observation he can fully confirm the evidence of Professor Forbes relative to the functions of the suckers of the Sea-Urchins. But although that talented author entertained no doubt as to the organs in question being powerful locomotive agents, he evidently seems to have felt himself unable to suggest any purpose they could possibly serve when situated on the back or upper part of the animal.

My own experience incontestibly proves that the suckers in question are used for precisely the same purpose as those situated in any other part of the body. I am enabled to state, from having repeatedly witnessed the phenomenon, that *the Echinus can walk about with equal facility while lying on its back as in its more natural position.* The advantage of this power to the animal under certain circumstances will be apparent upon a little reflection.

With regard to the spines, I fancy their purpose is almost solely to assist the Urchin to burrow in the sand, and to protect it from the attacks of its enemies. It may be, however, that at particular times they serve as aids to locomotion, but that their assistance can be, and is, often dispensed with entirely by the animal, I can most positively assert.

My experiments were always conducted in glass vases, up the smooth, polished sides of which my specimens frequently advanced. Upon reaching the surface of the water, I have seen an Urchin roll completely round and move along on its back, then after a time change its position, and travel round the circumference of the vessel *while attached by its side*, the body of the animal being sometimes inverted.

At such times as these it must be quite evident that the spines would be totally useless, and that by the suckers alone did the animal perform its interesting movements.

According to a certain writer, there are some foreign species of the Echini remarkable for possessing spines, which act both as offensive and defensive weapons. 'On one occasion' (this writer says) 'when searching for a fish in the crevice of a coral rock, I felt a severe pain in my hand, and upon withdrawing it, found my fingers covered with slender spines, evidently those of the Echinus, of a grey colour, elegantly banded with black.

'They projected from my fingers like well-planted arrows from a target, and their points being barbed could not be removed, but remained for some weeks imbedded as black specks in the skin. Its concealed situation did not permit me to examine this particular Echinus. In some experiments I approached the spines with so much caution, that had they been the most finely pointed needles in a fixed state no injury could have been received from them, yet their points were always stuck into my hand rapidly and severely.'

In addition to those above described, the Sea-Urchin is provided with other organs, in shape somewhat resembling minute pincers, supported on fleshy stems, which always keep up an incessant motion when the animal is in a healthy condition. They are scattered in great numbers over the surface of the body, among the spines, and around the mouth of the Urchin.

The use of these singular objects—by naturalists

termed Pedicellariæ — is totally unknown. Some writers think they are an integral part of the Echinus, others describe them as distinct and parasitic animals. There is good reason to believe that the former will eventually be proved to be the correct explanation of the matter.

Its masticatory apparatus is not the least wonderful portion of the Sea-Urchin. The teeth, five in number, which may frequently be seen protruding from the mouth, are of extreme hardness, and of seemingly disproportionate length. They are not fixed in sockets as ours are, or they would be speedily worn away by their action on the shelled mollusca upon which the animal feeds, but fresh substance is added to each tooth as fast as it is worn away by use, as in the case of many gnawing animals. ' In order to allow of such an arrangement, as well as to provide for the movements of the teeth, jaws are provided, which are situated in the interior of the shell, and these jaws, from their great complexity and unique structure, form perhaps the most admirable masticating instrument met with in the animal kingdom. The entire apparatus removed from the shell consists of the following parts. There are five long teeth, each of which is enclosed in a triangular bony piece, that for the sake of brevity we will call jaws. The five jaws are united together by various muscles, so as to form a pentagonal pyramid, having its apex in contact with the oval orifice of the shell, while its base is connected with several bony levers by means of numerous muscles provided for the movements of the whole. When the five jaws are fixed together in their natural position, they form a five-sided conical mass, aptly enough compared by Aristotle to a lantern, and not unfrequently described by modern writers under the name of "the lantern of Aristotle." The whole of this complicated machinery is suspended by muscles from a frame-work fixed in the interior of the shell, and may often be picked up upon the beach, or still better exposed *in situ* in a dead Echinus, by those who would examine closely this wonderful piece of mechanism.'[1]

I have made two careful drawings of the jaws and teeth of the Echinus. No. 1 represents, as it were, the 'elevation' of the pentagonal pyramid above described, while No. 2 constitutes the 'plan' of the same object.

The *Echinus sphæra*, or common Egg-Urchin, may often be seen forming a curious ornament in the drawing-rooms of the "West End," and also in the dwellings of the poorer classes, who, according to some authors, boil it like eggs, and so eat it. Hence its popular title. Among the ancients the Echinidæ were accounted a favourite dish. 'They were dressed with vinegar, honied

---

[1] The 'Aquarian Naturalist,' p. 224.

wine or mead, parsley and mint. They were the first dish in the famous supper of Lentulus when he was made Flamen Martialis. By some of the concomitant dishes they seemed designed as a whet for the second course to the holy personages, priests and vestals, invited on the occasion.'

The illustration on Plate X. was drawn from a living specimen, and gives a somewhat unusual representation of a Sea-Urchin. In general the spines alone are shown, but I have endeavoured to give the uninitiated reader some faint notion of the appearance which the *suckers* present when extended from the surface of the shell.

The young Urchin sat very quietly while I was engaged in taking his portrait, but continually extended crowds of his slender tubular legs in all directions, as above indicated, much to my gratification and apparently to his own.

In preparing a Sea-Urchin for a chimney ornament, the most important point is to remove the spines so as to let the tubercles remain entire. In performing this operation some little experience is necessary. Several times I attempted the process by aid of a pen-knife and a pair of pliers, but not with a satisfactory result. Having mentioned my difficulty to a friend, he laughingly asked me if I had ever heard of a certain pilgrim who, for some peccadillo he had committed, was doomed to perform penance by walking to Loretto's shrine with peas in his shoes?

Of course I was acquainted with the story, but could not see what it had to do with Sea-Urchins, and told my brother naturalist so. Still smiling, he said, 'Do you remember the relief that was said to be afforded to the humorous rascal, both mentally and bodily, by *boiling his peas?*' Yes. 'Well, then,' was the reply, 'do you boil your Sea-Eggs, and you will find your troubles speedily cease.' I did as I was directed, and found the advice of great service; for, after being an hour or two in the 'pot,' the spines of the Urchin may be totally rubbed off by aid of a nail brush, or some such instrument. Moreover, the colour of the shell is improved, and the dental apparatus may be drawn out entire, with the greatest ease.

I may here take opportunity to mention, that the student who may think proper to act upon the hint above given, should not boil the Urchin too long, or the fleshy parts will become dissolved, and the entire shell fall into a multitude of fragments.

This unfortunate result actually happened on one occasion to a genial, clever friend of mine, much to his chagrin and my malicious delight.

There are several other species of Sea-Urchins whose forms are tolerably well distinguished by their popular appellations. Thus we have the 'Silky Spined Urchin;' the 'Green Pea-Urchin,'—the latter is the commonest and prettiest of all its kindred, its back being covered with a kind of powdery green, as is seen on the elytra of many beetles; the 'Cake-Urchin,' which from its flattened form may be regarded as a link between the Sea-Urchins and the true Star-fishes; the 'Purple Heart-Urchin,' and the pretty 'Rosy Heart-Urchin,' appropriately named from the brilliant crimson hue that its body presents during life.

CHAPTER XX.
Sea-Cucumbers.
(HOLOTHURIADÆ.)

XX.

THERE is a very singular group of animals, the *Holothuriadæ*, that claims a passing notice, from their near relation in structural formation to the Sea-Urchins, although externally they also exhibit a certain resemblance to the *Annelides*. They are commonly termed Sea-Cucumbers, from the fancied likeness which they bear, both in shape and colour, to their namesakes of the vegetable kingdom.

A Holothuria is very unattractive in appearance when lying listless upon the sea-beach, but if a small specimen be transferred to the aquarium, it exhibits features of a very singular and interesting character. When about to change its position, the head, hitherto concealed, is protruded and expanded, until it assumes the form of a beautiful flower.

The animal moves principally by aid of sucker-like feet, similar in form to those of the Asteriadæ, or Sea-Urchins. In most species, the body is divided longitudinally into five rows of suckers. In some, however, these organs are scattered over the entire surface, while in the small Sea-Cucumber (*Psolus phantapus*), they are arranged in three rows upon a soft, oblong, flat disc, situated beneath the body of the animal, like the foot of a gasteropod mollusc.

Of one genus—the Trepang—many species are eaten by the omnivorous inhabitants of the Celestial Empire, by whom it is employed in the preparation of 'nutritious soups, in common with an esculent sea-weed, shark's fins, edible birds' nests, and other materials affording much jelly.' The intestines, which are generally found to be filled with coral, and solid masses of madreporic rock, are extracted, and the animal then boiled in sea water and dried in smoke.

Nothing can possibly be less enticing than the black and shrivelled carcases of these defunct gasteropods, as they are seen spread out and exposed for sale in the China markets. There are many varieties of Trepang, some being held in higher esteem than others,—hence the great difference which exists in the price of the article. The lowest quality being ten dollars, and the highest fifty dollars, per pecul of 133 lbs.

The following are titles by which a few of the Holothuriæ are known in China:—

    Great Black-Stone Trepang;
    Peach-blossom Trepang;
    Great White-Stone Trepang;

The Bald Trepang;
The Scarlet Trepang;
Great Clear-Ball Trepang;
The Middle Ash-Bald Trepang, &c., &c.

The illustration on Plate XI. gives a good idea of the typical form of the Holothuriadæ. It represents a species of the genus Cucumaria, *C. communis*, or common Sea-Cucumber. Its length is from four to eight inches; but, like all its kindred, it possesses the power of considerably extending or contracting its body at will. The Tentacula are ten in number, pinnate and plumose, stalked and rather large. The body is five-sided, with numerous suckers on the angles, but more on the sides, which are papilose. The colour is yellow, or brownish-white, although specimens found on the Irish coast exhibit a purplish hue.

This, the most common species of its genus, is an inhabitant of deep water, and is therefore most frequently taken with the dredge. Occasionally, specimens may be found after violent storms stranded on various parts of the shores of the United Kingdom.

The Sea-Cucumbers possess the singular power of disembowelling themselves upon the slightest provocation, and also of throwing off their Tentacula entire. There is one species, indeed, that exhibits a still more wonderful phenomenon. At certain times members of this species will divide their body into a number of parts, each of which will in due course become a new and completely-formed animal. After this the reader will be prepared to learn, that to build up a new inside, or create a new set of branchiæ, is to a Holothuria a very trifling and insignificant task.

CHAPTER XXI.

The Aplysia, or Sea-Hare.

'The origin and the source of the smallest portion of the universe overpowers our comprehension. How little can the acutest senses, the profoundest judgment, the widest view, embrace! It is as nothing; it is as less than nothing. We are capable of doing no more than surveying the edifice and adoring the Architect.'

SIR J. DALYELL.

XXI.

At several parts of the Scottish coast, and especially at North Berwick, may be found specimens of that curious gasteropod named the Aplysia, or Sea-Hare, the *Lepus marinus* of the ancients.

On visiting North Berwick during summer, I have been astonished to discover, in almost every pool, from two to twenty of these creatures.

At rest, the Aplysia is not by any means inviting, but when in motion, elevating and depressing the fleshy mantle that covers over the fringed and lobed branchiæ, its appearance is exceedingly graceful.

Striding across a pool on the look-out for some Gobies, whose forms darting beneath a large stone had not escaped my glance, I perceived the water in the rocky basin gradually lose its crystal brightness, and become changed to crimson. The Gobies were therefore allowed to rest in peace, while I proceeded to investigate a phenomenon that, at the moment, seemed somewhat singular.

A kind friend and brother zoologist, who happened

to be near, called attention to the fact that the crimson stream flowed thickest near where my foot rested.

On closely examining the spot pointed out, and turning over some fronds of Dulse, we came upon a small fleshy ball of a dark brown colour, from which there still issued a fluid of vivid crimson hue. Having placed this strange object in a bottle, I soon pronounced it to be an Aplysia, with whose full-length portrait, as represented in books, I had previously been made acquainted.

The power which this animal possesses, under irritation, of spurting out a peculiar secretion, I also remembered to have seen mentioned by several writers on natural history.

Although generally believed to be gentle and perfectly harmless, yet, as Professor Forbes observes, few molluscs have had a worse character than the Aplysiæ. From very ancient times they have been regarded with horror and suspicion; and many writers on natural history, conversant with them only through the silly stories of ignorant fishermen, have combined to hold them up as objects of detestation. To touch them, according to European prejudices, was sufficient to generate disease in the foolhardy experimenter; while Asiatics, reversing the consequences, maintained, perhaps with greater truth, that they met with instantaneous death when handled by man. Physicians wrote treatises on the effects of their poison, and discussed the remedies best adapted to neutralize it. Conspirators brewed nauseous beverages from their slimy bodies, and administered the potion confident of its deadly powers. Every nation in the world on whose shores the poor Sea-Hares crawled, accorded to them the attributes of ferocity and malignant virulence, although there never appears to have been the slightest foundation for a belief in their crimes.

A specimen of the **Aplysia that I** had in my tank deposited a stringy coil of spawn, which closely resembled that of the Eolis, with the exception that **the eggs,** instead of being white, were of a reddish **tint.**

CHAPTER XXII.
Serpulæ and Sabellæ.

XXII.

WITH the exception of the Balani (Acorn-Barnacles), perhaps the most common objects to be met with at the sea-shore are the Serpulæ. Scarcely a rock, or shell, or bit of old china, or piece of wood, or rusty nail, lying near low-water mark, but is encrusted with colonies of these animals. I have a small twig of a tree by me, so thickly coated with Serpulæ as to obscure all signs of its ligneous character, except at each end. A shell also exhibits the same phenomenon, and well-nigh defies the most skilful observer to define its original form with any degree of certainty.

The shelly tubes of these animals are built in the form of serpents, or twisted funnels, of a milk-white colour. Although so extremely hard, these tubes are formed solely by an exudation from the body of the animal—a simple marine worm. Unlike its erratic friend, the earth-worm, the Serpula is sedentary in its habits, and at no time does it ever leave its dwelling.

The delicate, but brilliant feathery plume—the only portion of the animal ever visible—constitutes the principal mechanism by means of which the Serpula constructs its calcareous tube.

A most wonderful instance of how mighty are the works which these insignificant creatures form when congregated together in vast numbers, and how useful such labours may sometimes be to mankind, is narrated by Dr. Darwin in his 'Voyage of the Beagle.'

Being delayed by adverse winds, this gentleman made a stay at Pernambuco, a large city on the coast of Brazil, and the most curious object that he saw there was the reef that formed the harbour. 'I doubt,' to use his own words, 'whether in the whole world any other natural structure has so artificial an appearance. It runs for a length of several miles in an absolutely straight line, and parallel to, and not far distant from the shore. It varies in width from thirty to sixty yards, and its surface is level and smooth; it is composed of obscurely stratified hard sandstone. At high water the waves break over it; at low water its summit is left dry, and it might then be mistaken for a breakwater erected by Cyclopean workmen. On this coast the currents of the sea tend to throw up in front of the land long spits and bars of loose sand, and on one of these the town of Pernambuco stands. In former times a long spit of this nature seems to have become consolidated by the percolation of calcareous matter, and afterwards to have been gradually upheaved, the outer and loose parts during the process having been worn away by the action of the sea, and the solid nucleus left as we now see it. Although night and day the waves of the open Atlantic, turbid with sediment, are driven against the steep outside edges of this wall of stone, yet the oldest pilots know of no tradition of any change in its appearance. This durability is by far the most curious fact in its history; *it is due to a tough layer, a few inches thick, of calcareous matter, wholly formed by the successive growth and death of the small shells of Serpulæ, together with some few Barnacles, &c.* These insignificant organic beings, especially the Serpulæ, have done good service to the people of Pernambuco, for without their protective aid the bar of sandstone would inevitably have been long ago worn away, and without the bar there would have been no harbour.'

Nothing whatever appears to be known relative to the mode of reproduction of these Annelids. I have paid much attention to the subject, but as yet have not gained any positive information regarding it. The only fact which I consider worthy of being chronicled is the following: On one occasion, when quite a novice in Marine Zoology, while observing a beautiful group of Serpulæ seated on a stone, I saw issuing from out one of the tubes a kind of very fine dust, of a rich crimson hue, which continued to arise for nearly an hour in spite of repeated efforts to disperse it by aid of a camel hair pencil. At first I believed the 'dust' to be the 'remains' of a deceased serpula, but afterwards found that such was not the case, the annelid being alive and healthy. Never having seen the phenomenon since, it has been a great source of regret to me that I did not endeavour to discover what the dust was composed of; but have little doubt that the microscope would have shown it to be, in reality, the ova of the Serpula.

Another class of Annelidans, termed Sabellæ, like the Serpulæ, also build habitations for themselves, but not of the same materials. Instead of being white, the tubes of the first mentioned animals are brown in colour, and composed of minute granules of sand, or small shells, and lined internally with a gelatinous substance exuded from the body of the worm. On the interior of the oyster and other shells, and even in univalves occupied by the Lobster Crab, various tubes of Sabellæ may often be seen. They are, however, generally discovered congregated together, forming a kind of honeycomb mass in the fissures of rocks, or against the sides of rock-pools, or on the surface of small stones, &c.

A mass of Sabellæ tubes forms by no means an inappropriate or unpleasant object for the tank, as the animals are hardy, and will live for many months if the water be kept pure. Moreover, while in confinement, they do not live in luxurious indolence, but

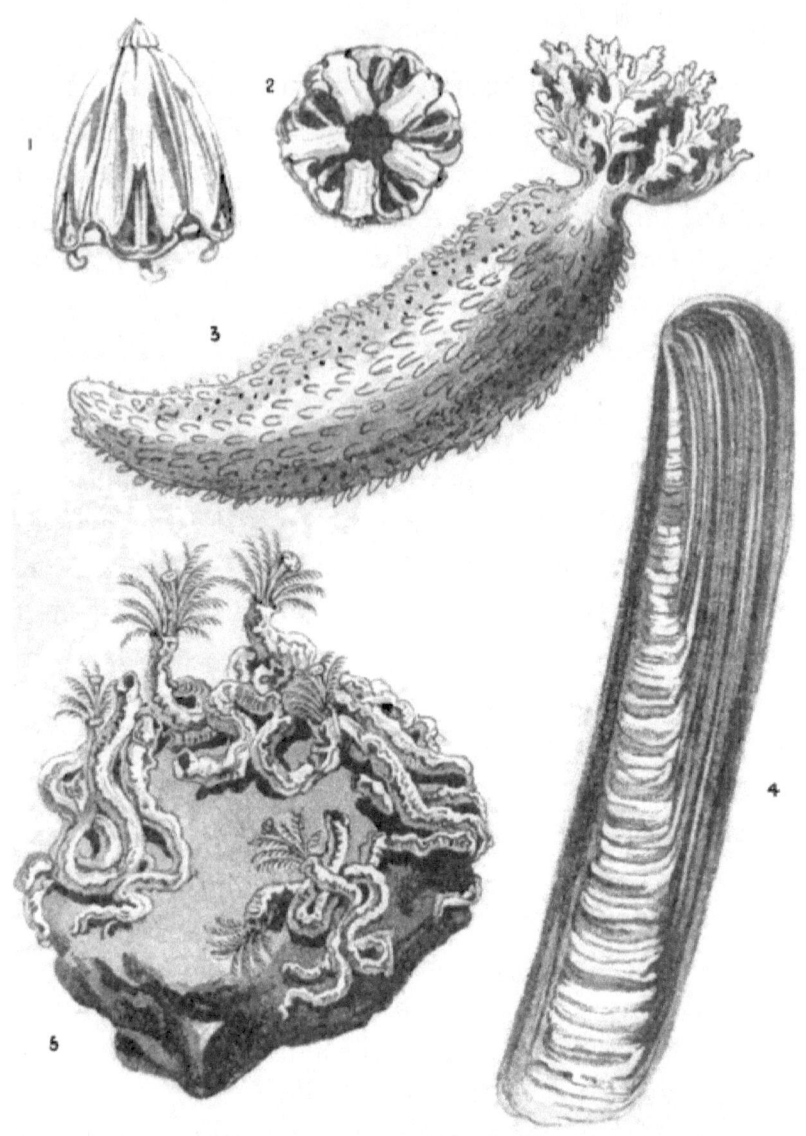

1, 2 SEA-URCHIN'S TEETH (Two illustrations.) 3 COMMON SEA CUCUMBER
4 COMMON RAZOR-SHELL. 5 COMMON SERPULÆ, attached to a piece of stone

ever seem to be busy in the exercise of their architectural propensities, making alterations, repairing damages, or otherwise 'sorting' their tubiculous habitations.

'The tubes of the Sabellæ,' says Dr. Williams, 'are soft, flexible, and muddy. Slimy mucus furnished by the integumentary glands of the body is the mortar or cement, fine sand molecules are the " stones " or solid material of the architecture. In the Sabellæ the lime of which the tubes are built is held in solution in the mucus provided by the cutaneous glands. It is adjusted in the fluid form, and moulded by appropriate tools into the required shape. It then *solidifies, too, under water,* like the " Aberthaw lime." The tube of the Sabellæ fits closely round the body of the worm; it is slightly elastic, and the interior is smooth.'

CHAPTER XXIII.

The Solen, or Razor-Fish.

'His mansion he extends,
So well concealed beneath the crumbling sands.

XXIII.

Few people who are in the habit of visiting the sea shore but must have noticed the empty shells of the animal about to be described. I allude to the Solen, or Razor-Shell, commonly so called from its resemblance to the handle attached to a barber's scythe.

This bivalve, improbable as the statement will appear to the uninitiated, is one of the most efficient burrowers to be met with on our shores.

By means of its fleshy foot it digs a hole in the mud or sand. Sometimes it retreats from the surface to a distance of several feet, but generally remains sufficiently near to allow its short, fringed siphons to project above the sand.

In walking along the beach, left bare by the receding tide, the pedestrian may often perceive little jets of water thrown up at his approach. These jets proceed from the Razor-Fish in question. Although we may be several yards from his burrow, his sense of feeling is so acute, that the faintest vibration of the earth around causes the creature to retire alarmed within his dwelling.

In many places the Solen is much sought after by the poor, who esteem it a great luxury. In foreign countries—Japan, for example—it is so highly prized that we are told, 'by express order of the prince of that country, it is forbid to fish them until a sufficient quantity hath been provided for the emperor's table.'

The Irish people, when they go out to catch the Solen siliqua, have an appropriate song and chorus which they sing, but whether to amuse themselves or charm the fish 'this deponent sayeth not,' for very obvious reasons. In general, I should think the less noise the more likelihood of success to those endeavouring to capture this animal.

'Who has not seen the picture of the stupid-looking boy going warily out with a box of salt, having been gravely informed by some village wag that if he would only just drop a pinch of salt on the birds' tails he would be sure to catch them. We are all familiar enough with this venerable joke, but not so with its successful application in another case. This time it is the fisherman, instead of the village boy, who carries the box. He cautiously slips a little salt into the hole, which irritates the ends of the siphons, and makes the *Solen* come quickly out to see what is the matter, and clear itself of this painful intrusion. The fisher, on the alert, must quickly seize his prey, or else it will dart back again into its retreat, whence no amount of salting or coaxing will bring it out again.'

If after reading the above quotation any person should fancy that in his mind's eye he perceives at many sea-side places, scores of hardy, weather-beaten fishermen walking about, each armed with nothing but a box filled with salt, wherewith to bamboozle the Spout-Fish, he will be most lamentably deceived. True it is, this plan is sometimes adopted by children and amateur naturalists, but by fishermen— never. Instead of a salt-box, these, when in search of their favourite bait, always carry a kind of harpoon, formed of a piece of iron rod, the end of which is sharpened to a point.

Having witnessed the Solen throw up his jet of water, and retire beneath the soil, the fisherman suddenly plunges his instrument into the orifice. Should the action have been skilfully performed, the rod will have pierced the animal between its valves, which instantly retract upon the intruding object. To draw the fish to the surface is then a comparatively easy task. If the first plunge of the rod be not successful, the fisher knows full well it would be futile for him to repeat the attempt, as the object of his attack would quickly burrow itself down to such a depth as to render pursuit hopeless.

Juveniles at the sea-side, imitating the plan above described, become by practice very expert in procuring specimens of the Razor-Fish by means of a piece of wire sharpened at one end.

CHAPTER XXIV.

A Gossip on Fishes &c.,

INCLUDING THE ROCKLING, SMOOTH BLENNY, GUNNEL-FISH, GOBY, ETC.

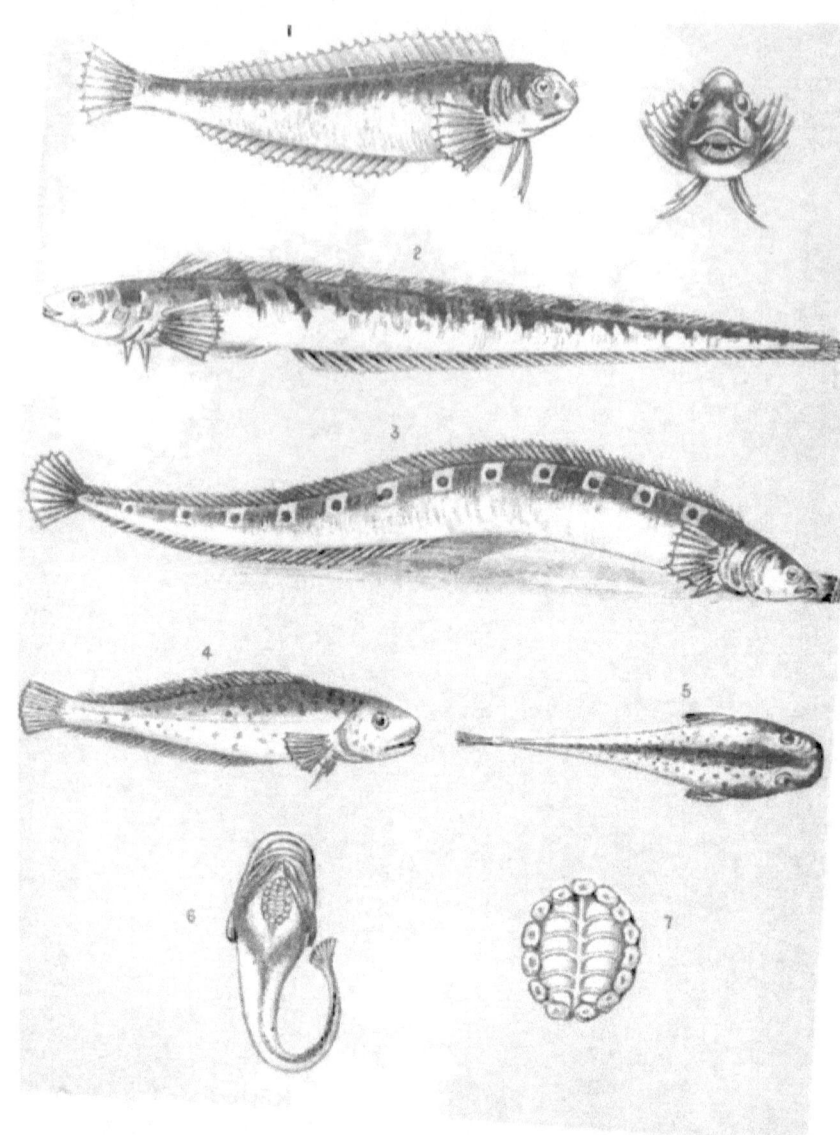

1 SMOOTH BLENNY
2 VIVIPAROUS BLENNY
3 SPOTTED BLENNY or Gunnel-Fish.
4.5.6 THE MONTAGU SUCKER FISH
(Three illustrations)
7 SUCKER of Dº

XXIV.

One of the best *bons mots* that I ever remember to have read was entitled, 'Punch's Address to the Ocean '—

'With all thy faults I love thee *still.*'

Any landsman who finds himself occupying a seat in a fishing-smack or oyster-boat while a stiffish breeze is blowing will, I am sure, with great mental fervour echo the above sentiment.

For myself, I can never take even a short trip on the water without experiencing some unpleasantness—proving to me that the sea is not 'my element.' Still, I am one of those to whom the 'salt ocean' is endeared by early recollections, having been, when a child, frequently among the aged and mutilated veterans of our country who vegetate on the banks of the 'silver Thames.'

From the tobacco-stained mouths of some of these old blue-jackets (all of whom, I may mention, according to their own account, had fought 'alongside of the *galyant* Nelson'), many strange stories have been poured into my eager and willing ears, and even now a thrill of delight is evoked when any of these 'yarns' rise to remembrance. Still, the truth must be told: ever since I narrowly escaped drowning by plumping into the water backwards, from leaning against the *unsnibbed*-door of a bathing-machine, and at another time from being in a boat that, to my intense horror and dismay, had sprung a leak— I have enjoyed the sea best when my feet are on dry land; in other words, I like to view the 'world of fluid matter,' in its various phases, from a distant and perfectly safe point of view. Nay more, I can always better appreciate certain of its beauties (at all events during winter time) when seated by a warm fireside.

When lately in such a cosy position, my thoughts reverted to the marvellous operations ever going on within the liquid walls of the great deep. There artifices and stratagems, robbery and murder, and cannibalism in its worst forms continually occur. On the other hand, there may be scenes of courtship, touching instances of maternal affection, such as, were they chronicled, would make our hearts bleed with truest sympathy. Still, the Rob Roy maxim of

'They should take who have the power,
And they should keep who can,'

seems therein to be carried out with a rigour that would do honour to the 'bold outlaw Macgregor.'

Might there is generally predominant over right. Fishes eternally prey upon each other; and for such reason, were it not for the wonderful fecundity of these creatures (one cod-fish, for instance, producing several millions of ova in a single season), we should soon have the waters depopulated of all but the monsters of the deep.

Now, knowing that such a state of things exists—that cannibalism is of such frequent occurrence, and the dogs of war are there ever let loose—the inquiry naturally presented itself: Are the inhabitants of the ocean a happy race or not? According to many writers, the answer must be given in the affirmative; nay, more, some authors state, and with good show of authority, too, be it observed, that *fishes are in reality the happiest of created things*, by reason that they have no fear or apprehension of death, nor are they subject to pain or disease, nor, in fact, to any of those ills that *flesh* is heir to. These creatures cannot, of course, live for ever; but by a merciful dispensation of Providence, their final pang endures but for an instant.

The celebrated St. Anthony is among the believers in the consummate happiness of the finny tribe. There is on record a discourse said to have been preached by him to an assembly of fish, in which they are flattered to an amazing extent. It almost rouses one's jealous ire to find such fulsome adulation bestowed upon the lower animals, at the expense of all other objects in nature, not excepting *man* himself. There is, however, such a singular force and truthfulness in some of the expressions and sentiments which occur in the Jesuitical discourse alluded to (given by Addison in his 'Remarks on Italy'), that I cannot resist the temptation of quoting a few of its most prominent passages.

We are told that St. Anthony, feeling annoyed at certain heretics not listening devoutly to his preaching, he determined to teach them a lesson; and for this purpose went down to the sea shore, and called the fishes together in the name of God, that they might hear his holy word. The fish soon swam towards the speaker in vast shoals, and, having ranged themselves, according to their several species, into a very beautiful congregation, were addressed just as if they had been rational creatures.

The sermon commences in the following words:—

'My dearly-beloved Fish,—Although the infinite power and goodness of God discovers itself in all the works of his creation, as in the heavens, in the sun, in the moon, and in the stars—in the lower world, in man, and in other perfect creatures,—nevertheless, the goodness of the divine Majesty shines out in you more eminently, and appears after a more particular manner, than in any other created beings.

'It is from God, my beloved fish, that you have received being, life, motion, and sense. It is he that has given you, in compliance with your natural inclinations, the whole world of waters for your habitation. It is he that has furnished it with lodgings, chambers, caverns, grottoes, *and such magnificent retirements as are not to be met with in the seats of kings* or *in the palaces of princes!*

'You have the water for your dwelling—a clear, transparent element, brighter than crystal; you can see from its deepest bottom everything that passes on its surface. You have the eyes of a lynx or of an Argus; you are guided by a secret and unerring principle, delighting in everything that may be beneficial to you, and avoiding everything that may be hurtful; you are carried on by a hidden instinct to preserve yourselves, and to propagate your species; you obey, in all your actions, works, and motions, the dictates and suggestions of nature, without the least repugnance or contradiction.

'The cold of winter and the heat of summer are alike incapable of molesting you. A serene or a clouded sky are indifferent to you. Let the earth abound in fruits or be cursed with scarcity, it has no influence on your welfare. You live secure in rains and thunders, lightnings and earthquakes; you have no concern in the blossoms of spring or in the glowings of summer, in the fruits of autumn or in the frosts of winter. You are not solicitous about hours or days, months or years, the variableness of the weather or the change of seasons.'

The saint still further 'butters his fish' by reminding them, among other things, that they were specially favoured by God at the time of the universal deluge, they being the only species of creatures that were insensible of the mischief that had laid waste the whole world! He then begs of them, as they are not provided with words, to make some sign of reverence; give some show of gratitude, according to the best of their capacities; express their thanks in the most becoming manner that they are able, and be not unmindful of all the benefits which the divine Majesty has bestowed upon them.

He had no sooner done speaking, but behold a miracle! The fish, as though they had been endued with reason, bowed down their heads with all the marks of a profound devotion, and then went joyously bobbing around with a kind of fondness, as in approval of what had been spoken by the blessed father, St. Anthony.

Many of the heretics, as a matter of course, were converted at beholding the miracle; and the polite and pious little fishes, having received his benediction, were dismissed by the saint.

Shakspeare authoritatively asserts that—

> 'Travellers ne'er do lie,
> Though fools at home condemn them.'

Here I beg to differ with the sweet Bard of Avon, who, I am sure, would have retracted his statement had he read the above fishy discourse, and also the following among many other strange anecdotes which are published regarding the 'denizens of the deep.'

An Eastern traveller tells us that, ' in a certain river whose waters flow from Mount Caucasus into the Euxine, there arrives every year a great quantity of fish.' This information not being particularly novel in regard to most rivers, will fail to excite surprise in the mind of the reader. A different result, however, will follow when he hears that, according to Abon-el-Cassim, 'The people cut off all the flesh on one side of those inhabitants of the deep, and let them go. Well, the year following,' as this veracious writer avers, 'the same creatures return and offer the other side, which they had preserved untouched; it is then discovered that new flesh has replaced the old!'

This account reminds us of the tale of the traveller who reported that he had seen a cabbage, under whose leaves a whole regiment of soldiers were sheltered from a shower of rain. Another, who was no traveller (but the wiser man), said he had passed by a place where there were four hundred braziers making a cauldron—two hundred within, and two hundred without beating the nails in. The traveller, asking for what use that huge cauldron was, he told him, 'Sir, it was to boil your cabbage!' A wittily severe, but deserved rebuke.

There are many other statements regarding fishes which, although curious, are, nevertheless, to a certain extent true.

The Chinese, for instance, who breed large quantities of the well-known gold-fish, call them, it is said, with a whistle to receive their food. Sir Joseph Banks used to collect his fish by sounding a small gong; and Carew, the historian of Cornwall, brought his grey Mullet together to be fed by making a noise with two sticks.

In spite of these accounts, there are many writers who affirm that *fishes do not possess the sense of hearing at all;* and certainly a belief that these creatures are gifted with such a faculty is not necessary, in my opinion, in order to explain the above-mentioned phenomenon.

At the fountains, in the gardens of Versailles, the writer has seen numbers of fishes flocking together, and anxiously waiting for the subscriptions of the visitors. Now, had a bell been rung, these animals, doubtless, would have appeared at the edge of the fountain as usual; but had the bell *not* been sounded, and any human figure been visible, they would have taken up the self-same position.

I have, at various times, kept packs of fishes (Blennies, &c.), and tamed them, so that each member would feed out of my hand. For some time I used to attract them to the side of the vessel in which they resided by striking a wine glass with a small stick; but I also noted that if I made myself visible, and remained silent, while handing down a few fish mouthfuls, that the whole pack followed as readily as if I had sounded the mimic gong. Nay, whether I offered any bribe or not, and silently approached their crystal abode, the whole family would immediately flock in great haste towards me.

The tameness of these little creatures was somewhat remarkable. On numberless occasions I have taken them up in the palm of my hand, without the slightest opposition on their part, and then stroked and smoothed them on the back, as I would do a bird. At such times they made a kind of musical chirp, expressive of pleasurable emotion, and seemed in no hurry to escape into their native element even when I laid my hand in the water.

Such delightful confidence was always rewarded with some dainty.

Dr. Warwick relates an instance of instinct and intelligence in the Pike, which is so remarkable that I am sure my readers will be pleased to be made acquainted with it. I am the more induced to transfer it to these pages, from the remarks with which the doctor closes his narrative. From reasons stated above, the reader will be prepared to learn that I do not consider the statements therein advanced—that fishes are really sensible to sound—by any means conclusive.

When residing at Dunham, the seat of the Earl of

Stamford and Warrington, he (Dr. Warwick), was walking one evening in the park, and came to a pond where fish intended for the table were temporarily kept. He took particular notice of a fine pike of about six pounds weight, which, when it observed him, darted hastily away. In so doing it struck its head against a tenterhook in a post (of which there were several in the pond, placed to prevent poaching), and, as it afterwards appeared, fractured its skull, and turned the optic nerve on one side. The agony evinced by the animal appeared most horrible. It rushed to the bottom, and boring its head into the mud, whirled itself round with such velocity that it was almost lost to sight for a short interval. It then plunged about the pond, and at length threw itself completely out of the water on to the bank. He (the doctor) went and examined it, and found that a very small portion of the brain was protruding from the fracture in the skull. He then carefully replaced this, and with a small silver toothpick raised the indented portion of the skull. The fish remained still for a short time, and he then put it again in the pond. It appeared at first a good deal relieved, but in a few minutes it again darted and plunged about until it threw itself out of the water a second time. A second time Dr. Warwick did what he could to relieve it, and again put it in the water. It continued for several times to throw itself out of the pond, and with the assistance of the keeper, the doctor at length made a kind of pillow for the fish, which was then left in the pond to its fate. Upon making his appearance at the pond on the following morning, the pike came towards him to the edge of the water, and actually laid its head upon his foot. The doctor thought this most extraordinary, but he examined the fish's skull and found it going on all right. He then walked backwards and forwards, along the edge of the pond for some time, and the fish continued to swim up and down, turning whenever he turned; but being blind on the wounded side of its skull, it always appeared agitated when it had that side toward the bank, as it could not then see its benefactor. On the next day he took some young friends down to see the fish, which came to him as usual, and at length he actually taught the pike to come to him at his whistle, and feed out of his hands. With other persons it continued as shy as fish usually are. He (Dr. Warwick) thought this a most remarkable instance of gratitude in a fish for a benefit received, and as it always came at his whistle, *it proved also what he had previously, with other naturalists, disbelieved, that fishes are sensible to sound.* (?)

On hunting among the rock-pools by the sea-shore, several peculiar little fishes are frequently to be found, and although some of them cannot be considered suitable for the aquarium, still, for the reader's information, it may be as well that I devote a brief space to a description of the peculiarities of each.

By far the most interesting of all the finny occupants of the rock-pool, is, to my taste, the Smooth Blenny, or, as it is variously termed, Shanny, or Tansy. It is also more abundant than many other species, and may therefore be readily captured during summer. The Blenny varies from two to five inches in length. The back is ornamented with exquisite markings, but the most characteristic features are the peculiar bluntness of the head, and the brilliant crimson dot both on and immediately beneath the eyes.

Although easily tamed, the Blenny, in his native haunts, appears to be the most timid of animals, darting with the rapidity of lightning to the shelter of some stone or overhanging weeds at the remotest indication of approaching footsteps, or the faintest shadow of a human form being cast on the water.

When desirous to procure a specimen, it is best to choose as small a pool as you can for your hunt. Drop in your net at one end, and as the Shanny precipitately retreats to the other, give him chase. Having arrived at the extremity of his domain, he will endeavour to hide among the weeds, but if you hold your net across the pool with one hand, and with the other lift up a stone or beat the bushes, the little fellow will become greatly excited, and darting out, of course, unwillingly, falls into the snare prepared for him.

Having gained your prize, do not handle it, but placing your finger under the net, tilt it over the mouth of the bottle, and allow the Blenny to fall as gently as possible into the water. You need be under no uneasiness after introducing him to the aquarium about the nature of his diet. He is far from being epicurean in his tastes. I supply mine according to my whim at the moment, with whatever is at hand, a bit of fowl, roast beef, or the like.

The only caution I adopt when giving animal food to the Blenny is to remove all traces of fat. I mince their food into minute particles, and having sufficiently moistened it, I place a morsel upon a hair pencil. This attention to their comforts the Blennies soon learn to appreciate, and will, after a while, display at meal times the sagacity of larger animals.

Perhaps the simplest plan to adopt is to cut open a mussel and throw it into the tank. A considerable deal of amusement, moreover, is often to be obtained by watching the fishes engaged at such a meal. How they toss the valves of the Mytilus about, and snap at each other's tails! How vexed they become if by accident the shelly dish is turned topsy-turvy, and resists all their manœuvres to reverse it so as to get at the meat! The valves of a large mussel will sometimes be literally cleaned out by some half dozen Blennies in the course of an hour.

I have noticed a singular fact in connection with the Blenny—namely, *that they do not all increase in size as they grow older.* Out of five that I kept domesticated for more than two years, one specimen remained at the end of that period of the same size as when I first made its acquaintance in a rock-pool by the sea-shore, while its companions had greatly increased their proportions. But let me in justice add, that if my little finny pet failed to increase in corpulency, it gained largely in intelligence. Who is there that has not seen children, short in stature, and comparatively old in years, who deserve the epithet applied to them by the vulgar, of 'little—but *knowing*.' This remark would apply with great truth to my 'little Dombey' fish.

Before becoming expert in carrying out the plan (which will be fully detailed hereafter) for clarifying the water of an aquarium which has become opaque from superabundant vegetative growth, I had to submit to many annoying failures. Thus it was in a certain instance.

I had cleaned out my tank, refilled it with partially purified water, and again inserted the various animals constituting my 'stock.' Emboldened by the success which had attended my operations, I thought a still further dose of diluted acid might be added, in order thoroughly to remove the greenish hue of the water. A few minutes showed me the folly of not letting well alone, for soon flakes of discharged vegetation were precipitated to the base of the vessel, covering it with a coating of fur.

The poor Blennies speedily showed signs of distress, and changed colour, as they generally do, upon the most trifling cause. Instead of dark brown or black, their bodies appeared of a yellowish tint, spotted with white. Such a change was lovely to the eye, but, alas! it was—

> 'The loveliness in death,
> Which parts not quite with parting breath.'

The little creatures jumped and dived about in all directions, all their motions being extremely violent. I quickly perceived the error which had been committed, and, moreover, discovered to my chagrin that such error could not possibly be rectified for some time, on account of my not having by me any reserve of pure salt water. Taking several of the fishes in my hand, I stroked their backs with a camel hair pencil, and was pleased to find that as their alarm subsided their natural hue returned. My being obliged to place my pets in their unhappy and pestilential home again was, as the reader may suppose, a source of regret to me; but I had some hopes that they might by chance survive, and become used to the 'vapour of their dungeon,' at all events until such time as I could hasten to the sea-side and procure a new supply of water. My expectations of such a result were built upon the fact, that although four of the fishes had changed colour, the small Blenny still retained its natural hue. How did this happen? it will be asked. I answer, by little Dombey (doing as his brethren had always hitherto done in similar circumstances) leaping on to a ledge of rock that projected out of the water, and there breathing the fresh air in safety.

On the following morning I peeped into the vessel, and saw by their upturned gills that all my finny proteges were dead!

> 'All my pretty ones?
> Did I say all?'

All except the smallest of the pack, he was still dressed in his sombre coat, and gracefully reclining upon the rocky couch above mentioned.

How thankfully he received the breakfast that I temptingly offered upon the tips of my feeding brush, and how grateful he seemed to be, when, after the lapse of a few hours, I was enabled to let him float again in his pure native element, a fresh supply of which had been procured with as little delay as possible!

The Viviparous Blenny differs from the other British Blennies 'in the circumstance to which its name refers—that of bringing forth its young alive, which seem perfectly able to provide for themselves from the moment they are excluded.'

It is a most gentle, graceful-looking fish, but as far as my experience goes, one that is impossible to tame, or rather, I should say, embolden. All my efforts to domesticate various specimens have proved unavailing; and in spite of the most earnest and kindly attention, they have generally pined away and died within a week after their introduction to the aquarium.

From the illustration on Plate XII. the reader will have no difficulty in recognising the original, should he by chance meet with it hiding among the tangle, or beneath the stones by the sea-shore.

The spotted Blenny, Butter-Fish, or Gunnel-Fish, as it is variously termed, is found lurking under stones in the same places as the preceding. In the north of Scotland it is called 'cloachs,' and is used extensively as a bait for larger fish. When disturbed, it wriggles its body about in the muddy bottom of the rock-pool like an eel, for which, indeed, it is occasionally mistaken.

Its length varies from three to nine inches; the depth only half an inch; the sides very much compressed and extremely thin.

The dorsal fin consists of seventy-eight short spiny rays, and runs the length of the back almost to the tail. The most conspicuous feature in the Gunnel-Fish are the eleven round spots which occur at the top of the back, and reach the lower half of the dorsal fin; they are black, half encircled by white.

The tail is rounded, and of a yellow colour. The back and sides are of a deep olive; the belly whitish.

In its young state I have had this fish live in my aquarium for several months, but it never seemed to be happy or contented.

The Five Bearded Rockling is almost as great a favourite with the writer as the Smooth Blenny. It is a very pretty fish, and may be easily tamed. In the course of a week I trained one to feed out of my hand, and when I put my finger in the water the fish would rub against it with its head, just as a favourite cat frequently does against the leg of a person with whom it is very familiar; moreover, if I moved the intruding digit with a circular motion through the water, the Rockling would waltz round the tip with evident signs of pleasure.

This fish is often found in tide-pools, and may readily be identified by the prominent appendages attached to its head, to the presence of which the Rockling owes its familiar appellation.

The Goby (*Gobius unipunctatus*), or, as it is more popularly termed, One-Spotted Goby, is frequently found inhabiting the same pool as the Blenny or the Rockling. The distinguishing character of this pretty creature is the black spot which is situated between the fifth and sixth ray of the first dorsal fin. Its length is usually about one, or one and a half inches; specimens have, however, sometimes been found on the shores of the Frith of Forth, that measured nearly three inches.

The colour of the Goby is very changeable. If the

animal is labouring under excitement, its body assumes a deep brownish tint, approaching in some instances to black; this gives place to brown, drab, and even amber, or yellowish white.

The Goby possesses the power of attaching its body to any object by means of its ventral fins, which become united together in the form of a funnel.

Another species (*Gobius bipunctatus*), or Two-Spotted Goby, is generally found among the *Fuci*, in rocky situations. Its name is derived from a dark spot which is distinctly apparent on each side, near to the origin of the pectoral fin.

The head and upper part of the body is dark brown, —the under part of the head and belly white or pale drab.

Allusion has already been made to the peculiarity of the Gobies affixing their bodies to rocks or other substances, by means of a sucker formed by the junction of the ventral fins. The adhesive power in question, which this class of creatures possess, is very limited as compared with that which is exercised by the true sucker fishes, and especially by the members of a certain species, whose bodies are furnished with two distinct organs of adhesion.

The extraordinary adhesive powers of the Lump-Sucker, for instance, have been tested by several writers. One observer states, that a fish of moderate size has been known to suspend a weight of above 20 lbs., upon which it had accidentally fastened itself. Mr. Pennant says still more, for he has known that, in flinging a fish of this kind just caught into a pail of water, it fixed itself so firmly to the bottom, that, on taking it by the tail, the pail was lifted up, though it contained several gallons of water.

To descend from the largest to the smallest species, we arrive at the Montague Sucker-Fish, or, as it is sometimes called, the Diminutive Sucker, one of the most interesting little creatures to be met with at the sea-shore. At the coast near Edinburgh I have met with many specimens, equally well in the spring or winter season, as during the summer months. At such locality this species may therefore be pronounced common; yet it is comparatively unknown to most 'collectors' in the neighbourhood. Many, indeed, contend that my designation is erroneous. But having taken considerable pains to satisfy my mind upon the subject, I have no hesitation whatever in stating that the little fish in question is identical with that of the Montague Sucker.

Donovan, in his 'Natural History of British Fishes,'[1] was the first to illustrate and publish an account of this *petite* gem of ocean. His figures are copied from drawings made by Colonel Montague, who also furnished the description of the specimen delineated. With the important exception of the

---

[1] This splendid work, which was published in five volumes, between the years 1802–8, contains 120 exquisite illustrations, all, *with the solitary exception, unfortunately, of the Montague Sucker Fish*, accurately drawn and coloured from living specimens, procured at vast trouble by the author.

sucker—an organ of adhesion which is very nearly correct—the general appearance of the Diminutive Sucker-Fish as figured, is not at all satisfactory. Perhaps this is not to be wondered at, when we remember that the specimen from which the sketches were taken was very small indeed. Moreover, it was diaphanous, and is depicted as being principally transparent, spotted, and tinged with pink.

The Diminutive Sucker, in its adult state, is said to be from two to three inches in length; consequently Colonel Montague's first specimen must have been an extremely young one.

The usual colour is deep orange, varied with minute dark spots. The under parts of the body and throat are of flesh colour; the centre of the sucker being faintly tinged with crimson.

I have seldom met with specimens measuring more than one, or one and a quarter inches. It is a marked peculiarity in this Sucker-Fish, that when adhering to any substance it has a constant habit of curving the tail towards the head. In such position it will remain motionless for several hours.

There is little difficulty in capturing the Montague Sucker in its native haunts. It does not possess the power of darting to and fro with the speed of the Blenny, or most other fishes, but progresses through the liquid element with a peculiar quivering motion.

It is not a fish that can be recommended for the aquarium. A fortnight to three weeks is the longest time that I have been able to keep a specimen alive; indeed, until I adopted the plan of allowing each little captive to remain quiet and undisturbed in a dark and shady place, death ensued in the course of one or two days.

My illustrations (Plate XII.) having been carefully drawn and coloured from a living specimen, the student will, I trust, find no difficulty in recognising the Diminutive Sucker, should he be so fortunate as to meet with it in a rocky pool.

'There is also a fish called the Sticklebag, a fish without scales, but hath his body fenced with several prickles. I know not where he dwells in winter, nor what he is good for in summer, but only to make sport for boys and *women anglers*.' Thus contemptuously does dear old Izaak speak of the Sticklebag, or Stickleback, as it is now termed, one of the most amusing and interesting members of the finny tribe. I have frequently transferred specimens of the Stickleback from fresh water to salt water, and found them live quite as well in the latter as in the former.

The contrast, however, between the appearance of the three spined Stickleback, when first taken from the sea, and one captured in the fresh water pond is very remarkable. The first is dressed in a gorgeous coat of varied colours. Around the mouth and belly it is bright crimson, on the upper part of its body various tints of green prevail; while in the pond specimen no red colour is visible at all, but only white blended with green.

In addition to *Gasterosteus aculeatus*, whom we have above alluded to, there is another species, *G. spinachia*, or Fifteen-Spined Stickleback, which is also an inhabitant of rock-pools by the sea-shore, but unlike the first-mentioned, is never found in fresh water. Both species possess one peculiarity in common, a description of which will form an appropriate conclusion to this chapter. I allude to their nest-building habits, which has only of late years been proved to exist, although Aristotle has recorded the same fact regarding a fish (*Phycis*) in the Mediterranean Sea, which was known to make a nest and deposit its spawn therein.

The duties of mason and architect are invariably undertaken by the male Stickleback. His materials are of course very limited, still his labours are skilfully and even artistically performed. Having chosen a suitable spot as a foundation for his house, he collects some delicate sea-weeds, gravel and sand, and with these materials, aided by a glutinous fluid which is given off from his body, the house is built. When completed, and not before, he seeks out his mate, and invites her to take possession of her newly formed home. If she shows any affection or coquetishness, he does not hesitate to nip hold of her tail, and urge her forward by equally expressive signs. Soon, like a dutiful little pet, she enters, and having deposited spawn, retires again, leaving her lord and master to guard the casket and its living treasure. This task, though extremely arduous, he adopts with pride and gratification.

How so small a creature can bear up so long under such a state of apparent excitement appears marvellous. His assiduity is most extraordinary. By night he rests beside the nest, and by day, if he can possibly hinder it, he allows nothing to approach. When there are other members of the Stickleback family in the aquarium, numerous combats are sure to ensue, for as the young and transparent offspring of one fish are deemed a great dainty by the non-parental body, the latter invariably endeavour to satisfy their cannibal propensities at the harrowing expense of their neighbours.

When the spawn are hatched, fresh care devolves upon the parent, in order to keep them within the nursery, and protect them from the greedy mouths of the larger fish, always on the look-out for tit-bits. Should one of the little fishlings stray beyond the prescribed bounds, the watchful parent darts after it, and in an instant his jaws close over the wanderer apparently for ever, but in fact only for a time, for swimming quickly back the old fish puffs out the straggler into its nest lively and uninjured.

CHAPTER XXV.

On the Formation of an Aquarium, &c.

'And so I end this little book, hoping, even praying that it may encourage a few more labourers to go forth into a vineyard which those who have toiled in it know to be full of ever fresh health, and wonder, and simple joy, and the presence and the glory of Him whose name is Love.'— C. KINGSLEY.

A SKETCH FROM NATURE.
1 Mussels attached by their byssus threads to the glass. 2 Fronds of *Chondrus chrispus*
3 Fronds of *Delesseria Sanguinea* 4 Fronds of *Ulva latissima*

XXV.

No ornament for the drawing-room or parlour can possibly be more beautiful than a well-stocked and tastefully-arranged aquarium; nor is there one likely to be productive of a greater amount of pleasure and amusement. And it is instructive as well as amusing, for by means of it the statements of writers relative to the habits of certain marine animals may be verified by personal observation, and even difficult problems in natural history satisfactorily solved. Aided by one of these 'mimic oceans,' let the reader commence the practical study of marine zoology, and I have little fear of his ever becoming tired of it.

> 'Age cannot wither it, nor custom stale
> Its infinite variety.'

When pursued even in the most humble way, this recreation yields a degree of interest greater than any other 'hobby' can produce, at least in an equal space of time. If engaged in business during the day, the student can always devote an hour morning or evening to the aquarium, and when least expected, some circumstance will take place to excite his wonder, and fill his mind with deep and devout reflection. Moreover, the young naturalist will undoubtedly derive pleasure from his endeavours to establish published facts relative to many of his little prisoners; pleasure in noting down any interesting anecdote that may occur; pleasure in knowing that his time is being profitably spent; and above all, that he is making himself acquainted with objects framed with marvellous skill and care by the hand of the Almighty:—

> 'Wonderful indeed are all His works,
> Pleasant to know, and worthiest to be all
> Had in remembrance, always with delight.'

Without further preface, I shall now proceed to offer some practical hints relative to the establishment of a marine aquarium. And, as some of my readers may be perfectly unacquainted with the subject, I shall treat it in as simple a style as possible. If, however, the experienced zoologist will kindly follow me to the end of the chapter, it may be that he will find some hints sufficiently new and useful to repay him for his trouble.

First, then, in regard to the tank. This indispensable requisite may be procured at certain shops in almost every town in the United Kingdom. Its price varies from two or three shillings to £20. The expensive kinds are generally oblong in form, but their construction being somewhat intricate, they are apt to get out of order, unless made by skilful and thoroughly competent artistes. Under certain circumstances, there is no doubt that an oblong tank of moderate dimensions is a great desideratum; but what I wish particularly to impress upon the mind of the reader is, that a large tank is not at all necessary in order to study the habits of marine animals; indeed, the more capacious the vessel, the more difficult becomes the task of watching the secret movements of any of its occupants. On this account it not unfrequently happens that a common glass tumbler becomes of much greater service to the student than the most elaborate aquarium.

The tanks which I use are circular in form, the largest being not more than sixteen inches in diameter, by seven inches in depth. Its cost was four shillings. Each one rests on a base of mahogany, elevated on turned legs to a height of nine inches.

Some persons object to the circular tank, on the ground that its occupants when seen from the sides appear magnified. This fact, as I have elsewhere remarked, is rather a recommendation with me, as it presents more distinct views of each movement in the vessel, and whenever I wish to see the objects of their natural size, I can do so by looking in from the top.

On the edge of the tank are placed three chips of gutta percha, in which are inserted three steel pins with brass heads; on these there is laid a circular piece of common glass, cut two inches larger than the diameter of the tank. As the 'pins' are about three-quarters of an inch above the tank, they allow a current of air to pass over the water, and also prevent, to a certain extent, particles of dust from falling in. On the edge of the movable lid I *paste* some crimson lace, which serves for ornament, and also prevents the glass from cutting the hand of any person moving it about. Sometimes I have a circular piece, about four inches in diameter, cut out of the centre of the glass lid, which allows the latter to be lifted off easily.

A glass syringe to aerate the water occasionally, a camel hair pencil, an ivory crotchet pin, and a pair of gutta percha forceps, complete the whole machinery of the aquarium, the cost of which is so trifling that the poorest person might manage to procure them.

One great point in favour of an aquarium, and one by no means generally understood is, that having once filled the tank with salt water, it will last for months, and even years, if proper care be taken, without requiring one particle of sea-water to be again added ; for as the water evaporates, the salt falls to the bottom, and the deficiency may be supplied with *fresh* water from the cistern or filter. In order to ascertain when the sea-water is of the proper density, you require to have a ' gravity bubble,' which can be had for sixpence. This may always be kept in the tank. When 'all's well' it sinks to the bottom, and when anything comes amiss it rises to the surface, but falls again quickly upon the introduction of the fresh water.[1]

A more simple plan is, to mark on the glass the height of the fluid when the tank is first filled, then as the water sinks, raise it again to its original level by means of fresh water.

Many persons decline starting an aquarium on account of the great difficulty of procuring a proper supply of sea-water. This objection, of course, can be offered only by those who happen to reside inland; but even these need not now be discouraged, for an ingenious plan has lately been devised for sending the commodity in question through the post!

Mr. Bolton, chemist, Holborn Bars, London, supplies, not sea-water, but 'marine salts for the instantaneous production of sea-water.' About six ounces is sufficient to make a gallon, by the application of *fresh* water. The saline material here alluded to, is not an artificial chemical compound, but is produced by the simple process of evaporating sea-water itself. Those individuals so fortunate as to possess a marine villa, or any other more humble residence at or near the sea-coast, have no occasion

---

[1] *Vide* author's "Sea-side and Aquarium."

to resort to the scheme above-mentioned for filling their tanks, a pure supply of sea-water being attainable with scarcely any trouble whatever. A stone jar should be kept for this purpose only, and care taken that the vessel is perfectly free from any smell, as that of spirits, dirty corks, or the like, as any such impurity would quickly spoil the water.

It may not be uninteresting to some of my readers to know, that in France an aquarium cannot be established with the same ease as in England. In the former country 'the whole contents of the sea itself is a contraband article,—that is, the contents of the salt sea of the English Channel or the Atlantic Ocean.' One writer tells us, that staying on the French coast, he kept sea-anemones alive in glasses, but was frequently warned by his friends to be careful how he fetched water from the sea, lest the custom-officers should interrupt him. 'My bottle,' to use the writer's own words, 'being very small, they let it pass, on the principle that the law does not care about extremest trifles; had it been a pailful, the case would have been different. A lady keeping a marine aquarium, explained her wants to the local head of the customs. He came and saw it —found it beautiful, and being a gentlemanly man, with some love for natural history, he gave a written order for the procuring of any reasonable quantity of water from the sea. Every time the needful element was brought from the shore, it was accompanied by its passport, as formally as if it had been a cask of wine, or a suspicious stranger. French salt sellers thus enjoy the height of protection; they are protected even from their colossal competitor, the sea!'

I do not know a prettier sight than that exhibited by a healthy aquarium on a fine summer's day; the effect of the sunshine upon it being to cause innumerable bubbles of oxygen—that look like balls of quicksilver—to form on every weed, shell, and smallest pebble. On looking through the transparent sides of the vessel, small particles hitherto resting on its base, may be seen slowly arising to the surface of the water, each buoyed up by a miniature gas balloon. The broad, ribbon like fronds of the ulva, from the selfsame cause, float upwards, and reflect a beauteous emerald hue upon all objects that lie beneath; while the glass bulb, placed in the tank as before stated, to denote the density of the water, at such a time belies its mission, and covered with numerous argent globules, mounts gracefully in companionship with the sea-weed, until shades of evening approach, when its buoyancy gradually subsides, and once more it falls to its original resting-place.

Wherever the above phenomenon is apparent, rest assured that the aquarium is in good condition. It is, in fact, to the oxygen thus given out by the plants and infant vegetation that the animals owe their existence. If no algæ were introduced, the water would become impure, and unless changed often, your little colony would surely die,—at least those of its members who were unable to rise above the fluid, and occasionally breathe the fresh air.

The secret herein involved, that animal and vegetable respirations counterbalance each other, has only of late years been discovered; yet it is apparent to any observing eye at the sea-shore; there we never meet with a rock-pool containing living animals, that is not more or less adorned with sea-weeds.

The green Lettuce Ulva, so abundant in rock-pools, the sea-grass, which covers almost every fixed object at the sea-shore, or the well-known dulse or Chondrus Crispus, form the only sea-weeds that it is necessary to introduce into an aquarium. In fact, one or two fronds of the Ulva Latissima alone, will answer perfectly well to purify the water of even a comparatively large tank. I have often been surprised to find how small a quantity of algæ was required for the purpose mentioned. After allowing a single frond to float for a few days in a tank, in which some sea-water was newly deposited, I took it out, and for an entire twelvemonth the water remained healthy and as clear as crystal.

The arrangement of the 'stock' of an aquarium is quite a matter of taste; perhaps no two persons adopt precisely the same plan. It may, therefore, be advisable, as this matter is so arbitrary, for the writer to state how his own tanks are mapped out, leaving it to his readers to imitate the arrangements, or adopt a style of their own as they may think proper.

At one time I used to make a grounding of sand, but this plan is not to be recommended, even though it be one highly approved of by several species of crabs, &c. White pebbles do very well, but I now prefer to cover the base of the tank with crushed shells, washed very clean.

The following is a sketch of one of my tanks as it at present stands:—

In the centre of the vessel is a *semi-circular arch*, formed of pure white Sicilian marble, which has to my eye a most pleasing appearance. Around it, and indeed over the entire floor of the tank, are strewn chippings of the same material as the centre piece itself.[1] From the arch, at certain intervals, hang various sized specimens of the *Mytilus edulis*, which have gradually advanced to their more or less elevated positions entirely by their own unaided exertions. Near hand a hardy *A. mesembryanthemum* has taken up his abode, and sits with ever expanded tentacles, motionless and happy. On either side of the Anemone is deposited a riband of Doris spawn, that undulates to and fro whenever by any chance the water is in the slightest degree disturbed. Se-

---

[1] The arch was cut from one of the waste pieces, of which there are always a large number, lying in a marble mason's yard, and cost but a few pence. The 'chippings' may be had in most cases for the trouble of carrying them away.

veral soldier crabs, of course, act as sentinels of the tank, and appear to be ever 'on duty,' marching about in all parts of their subaqueous habitation; while beneath the marble fragments repose, each with his 'weather eye' open, a small *Maia squinado*, two long-armed crabs, and a small *Carcinus mœnas*. On the sides of the vase rest a Limpet, a Trochus, and two fine Periwinkles, with skin of glossy blackness. The shells of either 'Buckie' is covered with myriads of quicksilver globules, that rest on the tips of the young and rising vegetation like dew upon the bladed grass. As I write, upon the inner surface of the water, like a fly upon the ceiling of a room, an Eolis and two pearly white Dorides lie idly floating in close companionship. Beneath them, upon the verge of the aperture of a large empty whelk shell, sits a pretty, cream-coloured Plumose Anemone (*A. dianthus*). On two blocks of stone repose several specimens of that mysterious animal the Pholas, who, by my unkindness, are thus made to become members of the marine 'houseless poor.' Several young specimens of these bivalves are seated in a piece of rock, and daily engaged in 'boring.' A stick of wood, formerly the slender twig of a tree, is thickly clustered with fairy-handed acorn barnacles and serpulæ, and being placed against the glass, the movements of these singularly beautiful creatures can be watched with ease. Then there are two Star-fishes, a pack of three little Blennies, and a Five-bearded

Rockling, whose singular movements I have previously alluded to. Against the arch some fronds of ulva are anchored, while at chosen spots specimens of delicate sea-weeds are also fixed—these rising up, and being magnified through the sides of the vase, have a pleasing effect, even to the eye of a child.

It is a pretty sight to watch the fishes glide under and around the marble arch, or throw themselves upon its highest point, there to enjoy the fresh air, and have a pleasant 'crack' together. This expression is literally correct, for the Blennies, when thus situated, usually make a kind of noise not inaptly expressed by snapping the nail of the thumb and finger together.

The foregoing animals which constitute the entire stock of one tank, are, I am proud to state, all in a healthy condition, and if we may judge by appearances, all contented and happy. It will be from no fault of mine if they do not long continue thus, and exhibit no signs of yearning for their native haunts by the sea-shore.

> 'Those gay watery grots—
> Small excavations on a rocky shore,
> That seem like fairy baths or mimic wells,
> Richly embossed with choicest weed and shells,
> As if her trinkets nature chose to hide
> Where nought invaded but the flowing tide.'

In another tank I have introduced as a centre object a fine piece of white coral, the higher branches of which rise above the surface of the water. The roughness of the coral seems to be much approved of by many of the animals, who are not slow to avail themselves of the facility thus afforded them of climbing and otherwise exercising their peculiar propensities. When purchasing coral, care must be taken to procure a specimen that has not undergone any cleaning process, for although such may be more pleasing to the eye, it is not so suitable for a 'centre piece' as the cream-coloured, and less expensive coral.

A third aquarium which I possess is fitted up in a somewhat novel style, which offers, for certain purposes, some slight advantages over others that I have seen employed. It can be adopted in almost any kind of tank; but the one under consideration is circular in form, and is, in fact, a bell-shaped inverted fern glass, the knob of which is sunk into a stand of wood supported on three legs.

The plan alluded to, which was suggested to the writer by an ingenious friend,[1] consists of the introduction of a floating centre piece composed of gutta percha, which serves as a resting place for various small animals, such as Actiniæ, Mussels, Barnacles, Serpulæ, and even Pholades and Cockles. At the base of the vessel, which is quite uncovered, rest sundry members of the crustaceous family, whilst

---

[1] Mr. Walter Hardie of Edinburgh, who has been my companion in many a delightful excursion among the rock-pools of the shores of the Frith of Forth, and to whom I feel myself greatly indebted for much valuable information relative to the subject of marine zoology.

fishes of various kinds swim freely about over the entire vessel free from all annoyance.

The question will doubtless be asked, 'How can I procure the centre piece here spoken of?' I answer, Make it yourself; a little skill combined with patience and gutta percha being all that is required. The following directions will serve to aid the young reader who may wish to test his manipulative powers.

Procure a thin piece of gutta percha, and lay it in hot water for a few minutes until it is thoroughly soft and pliable. Then get a globe—an orange will do if nothing better offers—and cover it with the above material. Having done this, throw it into cold water, and when hard, cut the fruit in two, so as to leave the gutta percha cast to the shape of each half.

Next make a circular tray about eight or nine inches in diameter, and turn up its edge about half an inch all round. Then heat the brim of each cup, and fasten them to the centre of the upper and under part of the 'tray.' The structure will then float in water. This, however, is not all that you want, as your centre piece must always be entirely immersed. First bore a few holes in the tray, then fix a pretty shell, with a hole in it, to the base of the lower 'cup,' and also form a loop of gutta percha, from which to suspend, by means of a piece of silk, a fragment of stone or marble of sufficient bulk to balance the centre piece, and sink it an inch or two below the surface of the water. At the centre of the upper cup fasten a small piece of gutta percha tube, at the end of which the valve of a Pecten may be attached as an ornament. The whole structure must be gently warmed and entirely coated with fine sand ; then tastefully decorated with shells and fronds of green Ulva, and the crimson Delesseria Sanguinea.

Sometimes I introduce a globe of glass as a buoy, and to its centre attach the tray of gutta percha.

A useful centre piece, a specimen of which I have had in use for several months, may be formed thus. Make a tripod of gutta percha, on the top of which attach the valve of a Pecten. From the centre of this object fix a branch of coral by aid of gutta percha, in such a way that it rises above the water in the tank. From under the shell pieces of coral may be made to branch out in various directions. The stand should be coated either with crushed shells or sand, to give it an ornamental appearance.

It is often a source of annoyance to find the base of the aquarium so thickly covered with dirt, &c. To get rid of this great 'eye sore,' without emptying and re-arranging the tank, I call in the aid of a very simple and effective instrument. By its application all objectionable matter may be gradually removed without in the slightest degree disturbing the water, or materially displacing the objects situated at the base of the vessel.

The instrument mentioned is composed of a gutta percha globe, made in the manner previously described, into one end of which is inserted a tube of gutta percha or glass about four inches long, and at the opposite end of the ball is introduced a second tube about eight inches in length.

To use this instrument, close the orifice of the longest tube, and plunge it into the water over any spot where the debris is collected, then by removing your finger from the end of the tube, *the impurity will be instantly sucked up into the ball*. By again placing the finger in its former position, the siphon may be lifted out of the tank, and its contents allowed to run off into a jug or basin placed near for the purpose.

This operation must be repeated until the whole of the offending particles are removed. Of course, more water will be drawn off than is necessary, but it can easily be poured back into the tank as soon as the sediment has been fully precipitated.

Aquaria are generally much more difficult to keep in order in summer than in winter, owing to the rapid and profuse growth of minute vegetation which renders the water opaque and exceedingly unpleasant to the eye.

This ugly opacity I at one time attributed to decaying animal matter, for I could scarcely believe that the mere increase of the algæ spores could produce such a vile effect. Experience, however, has proved that the latter was in reality the true cause. I tried often by syringing the water, or drawing it off by means of the siphon, or stirring it about in all manner of ways, to remove the objectionable muddiness, but always without success. Limpets and Periwinkles seemed quite useless. Nor did shutting out the rays of light for a few days have any perceptible effect in subduing the growth of the algæ which collected with wondrous rapidity, and arrayed each stone, shell, pebble, and even the poor crabs, in a greenish garb.

I was therefore under the necessity, on several occasions, of renewing the water, and considering that my residence was several miles from the sea-coast, this task was by no means a pleasant one. What made matters still more provoking, was the fact that the rejected fluid seemed perfectly free from all offensive smell. I now adopt the following novel method for removing the opacity of the water, without the latter being changed, and also for preventing the too abundant growth of the algæ at all seasons.

The plan in question (which requires, as already shown, to be carried out with extreme caution by the inexperienced aquarianist) is merely to dilute a small quantity of *alum* in a wine-glass full of water, and then mix it with the water contained in the aquarium. A pellet of alum about the size of a pea is sufficient for the purpose, if the tank be of moderate size. And if inserted on the first appearance of dimness in the water, much future trouble will be saved.

Supposing the water to have become opaque, proceed thus—Draw off a portion into a large jug, and mix with it the diluted acid as before stated, then let the jug remain undisturbed for about twelve hours. The vegetation having been deposited in flakes at the base of the vessel, the water should then be gently strained off through a piece of fine muslin into a second receptacle, which, in its turn, should be allowed to stand for some time, and the contents again strained as before. This process it is advisable to repeat several times, until the whole of the fluid in the tank has been thoroughly cleansed from impurity.

Should the water be returned too early, an unpleasant fur coating will appear over the entire base of the tank. This can be easily removed by means of the siphon. Let one end of this instrument, when in action, be passed gradually over the lower portion of the vase, and in the course of a few minutes every sign of 'fur' will be obliterated with the loss of but a small portion of water. When once the fluid has been clarified in the manner here mentioned, there is little fear of the young aquarianist being again troubled in like manner for many months, the acid apparently preventing the algæ from being reproduced to any such excess as hitherto.

GLOSSARY OF SCIENTIFIC TERMS.

Extracted principally from Professor Owen's learned work entitled, "Lectures on the Comparative Anatomy and Physiology of the Invertebrate Animals."

Ambulacra (L. *ambulacrum*, an avenue, or place for walking). The perforated series of plates in the shell of the Sea-star, or Sea-urchin, through which the sucking-feet are protruded.

Acalepha (Gr. *akalephe*, a nettle). The class of radiated animals with soft skins which have the power of stinging like a nettle. Commonly called Sea-nettles or Jelly-fish.

Actinæ Gr. *aktin*, a ray). The genus of Polypes which have many arms radiating from around the mouth.

Alternate generation. That modification of generation in which the young do not resemble the parent, but the grand-parent; so that the successive series of individuals seem to represent two species, alternately reproduced, in which also parthenogenesis alternates with the ordinary engendering by impregnation.

Algæ (sea-weeds). A large class of cryptogamic plants inhabiting salt and fresh water.

Anomoura (Gr. *anomos*, irregular, and *oura*, a tail). A section of crustaceous animals distinguished like the Hermit crabs, by the irregular form of the tails.

Annelid. The Anglicised singular of *annelleta*.

Adductor muscles, are those which hold together the shell of a bivalve, such as the Oyster, Mussel, &c.

Animalcules. Those extremely small animals which are invisible to the naked eye.

Antenna (from the Latin for yard-arm). Applied to the jointed feelers or horns upon the head of insects and crustacea.

Balanoids (Gr. *balanos*, an acorn). A family of Sessile cirripeds, the shells of which are commonly called Acorn-shells.

Bivalve. When a shell consists of two parts, closing like a double door. The mollusca so protected are commonly called bivalves, as the Mussel.

Brachyura (Gr. *brachus*, short; *oura*, tail). The tribe of crustacea with short tails, as the Crabs.

Branchiae. The gills or respiratory organs which extract the oxygen from air contained in water, as in fishes and other aquatic animals.

Buccal (L. *bucca*, mouth). Belonging to the mouth.

Byssus (Gr. *byssos*, fine flax). A term applied to the silken filaments or 'beard' of the Mussel and Pinna.

Carapace. The upper shell of the Crab, &c.

Calcareous. Composed more or less of lime.

Carnivorous (L. *caro*, flesh; *voro*, I devour). The animals which feed on flesh.

Caudal (L. *cauda*, the tail). Belonging to the tail.

Cephalópoda (Gr. *kephale*, a head; *pous*, a foot). The class of Molluscous animals in which long prehensile processes, or feet, project from the head, as in the Cuttle-fish.

Ciliogrades (L. *cilium*, an eyelash; *gradior*, I walk). The order of the *acalephæ* (as the Beröe) which swims by action of cilia.

Cilia (L. *cilium*, an eyelash). The microscopic hair-like bodies which cause, by their vibratile action, currents in the contiguous fluid, or a motion of the body to which they are attached.

Cirri (L. *cirrus*, a curl). The curled filamentary appendages, as at the feet of the Barnacles.

Cirripedes, or **Cirripedia** (L. *cirrus*, a curl; *pes*, a foot). A class of articulate animals having curled, jointed feet; sometimes written Cirrhipedia and Cirrhopoda.

Conchifera (L. *concha*, a shell; *fero*, I bear). Shell-fish; usually restricted to those with bivalve shells.

Comminuted. Broken or ground down into small pieces.

Conchology. The department of science which treats of shells.

Convoluted (L. *convolutus*). Rolled together.

Cornea (L. *corneus*, horny). The transparent horny membrane in front of the eye.

Crinoid (Gr. *krinon*, a lily; *eidos*, a discourse). A family of Star-fishes which bear some resemblance to the form of a lily. The fossils called Stone-lilies, or Encrinites, are examples.

Crustacea (L. *crusta*, a crust). The class of articulate animals (which includes the Crab, Lobster, &c.) with a hard skin or crust, which they cast periodically.

Decapoda (Gr. *deca*, ten; *pous*, a foot). The crustaceous and molluscous animals, which have ten feet, such as the Crab, Cray-fish, &c.

Digitate (L. *digitus*, a finger). When a part supports processes like fingers.

Effete. Barren, worn out.

Elytra (Gr. *elytron*, a sheath). The sheath or wing covers of coleopterous insects (Beetles).

Entomostraca (Gr. *entoma*, insects; *ostracon*, a shell). The order of small crustaceans, many of which are enclosed in an integument like a bivalve shell.

Entomology (Gr. *entoma*, insects; *logos*, a discourse). The branch of science treating of insects.

Exuvium, Pl. **exuviæ** (L. *exuo*, I cast off). The shell or skin of an animal which is shed in moulting.

Epizoa (Gr. *epi*, upon; *zoon*, an animal). The class of low organized parasitic crustaceans which live upon other animals.

Fissiparous (L. *fissus*, divided; *pario*, I produce). The multiplication of a species by the self-cleavage of the individual into two parts.

Frond (L. *frons*, a leaf). A term applied to that part of flowerless plants resembling true leaves.

Fucivorous (L. *fucus*, sea-weed; *voro*, I devour). Animals which subsist on sea-weed.

Flora. The plants which belong to a country or district.

Foliaceous (L. *folium*, a leaf). Shaped or arranged like leaves.

Gasteropoda (Gr. *gaster*, stomach; *pous*, a foot). That class of animals which (like the Snail) have the locomotive organ attached to the under part of the body.

Gemmiparous (L. *gemma*, a bud; *pario*, I produce). Propagation by the growth of the young like a bud from the parent.

Habitat. The locality in which an animal habitually resides.

Hinge. That part of a shell at which the valves cohere.

GLOSSARY OF SCIENTIFIC NAMES.

Hyaline (Gr. *hualos*, crystal). The pellucid substance which determines the spontaneous fission of cells.

Hydra (Gr. *hudra*, a water serpent). The modern generic name of certain fresh water polypes.

Hydrogen (Gr. *hydor*, water; *gemmæ*, I produce). A gas forming one of the components of water and atmospheric air.

Infusoria. The class of animalcules which abound in vegetable and animal infusions.

Lamellibranchiata (L. *lamella*, a plate; *branchiæ*, gills). The class of acephalous molluscs, with gills in the form of membraneous plates, of which the oyster and mussel are familiar examples.

Larva (L. *larva*, a mask). Applied to an insect in its first active state, which is generally different from, and, as it were, masks the ulterior form.

Ligament. A membrane close by the hinge which connects the valves.

Mantle. The external soft, contractile skin of the mollusca, which covers the viscera and a great part of the body like a cloak.

Macroura (Gr. *makros*, long; *oura*, a tail). A tribe of ten-footed crustacea (as the Lobster, Cray-fish), which have long tails.

Medusæ. A genus or family of soft radiated animals or Acalephæ, so called because their organs of motion and prehension are spread out like the snaky hair of the fabled medusa.

Molecules. Microscopic particles of matter.

Mollusc—Mollusca (L. *mollis*, soft). The primary division of the animal kingdom. It contains most shell-fish, slugs, &c.

Monograph (Gr. *monos*, one; *grapho*, I write). A written description of a single thing, or class of things.

Multivalve (L. *multus*, many; *valvæ*, folding doors). Shells composed of many pieces or valves, as the Chiton.

Nudibranchiate (L. *nudus*, naked; *branchiæ*, gills). An order of gasteropods, in which the gills are exposed, as the Eolis, Doris, &c.

Oxygen. A gas which is one of the constituent parts of water and of atmospheric air. It is essential to animal life.

Oviparous (L. *ovum*, an egg; *pario*, I bring forth). The animals which bring forth eggs.

Operculum (from the Latin for lid). Applied to the horny or shelly plate which closes certain univalve shells, as the Whelk, Periwinkle, &c.

Papillæ (L. *papilla*, a nipple). Soft prominences which resemble in form the teats of animals.

Palpi (L. *palpo*, I touch). The organs of touch commonly called 'feelers,' developed from the labium and maxillæ of insects.

Pectinated (L. *pecten*, a comb). Toothed like a comb.

Physograde (Gr. *physis*, air; *gradior*, I advance). The acalephes that swim by means of air-bladders.

Phytophagous (Gr. *phuton*, a plant; *phago*, I eat). Plant-eating animals.

Pulmonigrade (L. *pulmo*, a lung; *gradior*, I walk). The tribe of Medusæ which swim by contraction of the respiratory disc.

Rotifera (L. *rota*, a wheel; *fero*, I bear). The name of a class of infusorial animalcules, characterized by the vibratile and apparently rotating ciliary organs upon the heads.

Rhodospermes. The red-coloured seaweeds.

Serrated (L. *serra*, a saw). Toothed like a saw.

Sessile. Attached by a base.

Silicious (L. *silex*, a flint). Flinty.

Setæ. Bristles, or similar parts.

Spicula (L. *spiculum*, a point or dart). Fine-pointed bodies, like needles.

Tuberculate. Warty, or carved with small rounded knobs.

Testacea (L. *testa*, a shell). Molluscs with a shelly covering, as the Oyster, Whelk, &c.

Univalve (L. *unus*, one; *valvæ*, doors). A shell composed of one calcareous piece, as the Periwinkle.

Umbones. The base of a shell about the hinge.

Viviparous (L. *vivus*, alive; *pario*, I bring forth). The animals which bring forth their young alive. See Oviparous.

Whorl. The spiral turn of a shell.

Zoology (Gr. *zoon*, animal; *logos*, a discourse). That branch of science that treats of the habits, structure, and classification of animals.

Zoologist. One who is acquainted with the science of Zoology.

Zoophyte (Gr. *zoon*, an animal; *phyton*, a plant). The lowest primary division of the animal kingdom, which includes many animals that are fixed to the ground and have the form of plants.

INDEX.

Animalculæ, 37.
Actiniæ (Sea anemones), 38, 47.
    mesembryanthemum, 48, 365.
    troglodytes, 51, 62.
    bellis, 56.
    dianthus, 57, 62, 154, 366.
    crassicornis, 61, 100.
    coriacea, 56.
    parasitica, 56.
    explorator, 51.
Acorn barnacles, 145.
Adductor muscle, 178, 182.
Annelids, 154, 191, 315.
Acalephæ, 203.
Alternation of generations, 214.
Aphrodite aculeata, 267.
Aplysiæ, 54, 309, 311.
Aquariæ (on the formation of marine), 357.
Anomoura (Hermit crabs), 69, 92, 130, 133.
Asteriadæ, 271.
Algæ, 97.
A. ventilabrum, 161.
Amphitrite, 162.

Buccinum undutum, 92.
Byssus of Mussel, 168, 170, 177, 184.
Beröe, 210.
Bêches de mer, 31.
Brittle Star-fishes, 277.
Bird's foot Sea-star, 285.
Blenny (Smooth), **71**, 104, 236, 336, 341, 365.
Blenny (Viviparous), **346**.
Butter fish, 347.
Brachyura (crabs, &c.), 69, 133.
Buckie, 93.
Barnacles, 98, 146.
Barnacle geese, 150.
Boring Acephala, 251.
Bivalves, 122, 167, 282, 363.

Bearded rockling, 346.

Cilia, 35, 147, 214.
Coryne, 41.
Crabs, 67.
Cancer Pagurus (Edible crab), 67, 69, 128.
Carcinus maenas (Common Shore crab), 67, 78, 120, 127, 167, 239.
Common Whelk, 94.
Common Cockle, 84, 106, 239.
Cray fish, 128, 131.
Crangon vulgaris (Common Shrimp), 139.
Cestum veneris, 209.
Cydippe pileus, 210.
Cyanea capillata, 218.
Comatula rosacea, 275.
Crinoid Star fishes, 275.
Cross fish, 280.
Cushion stars, 285.
Cake Urchin, 300.
Chondrus crispus (Irish moss), 101, 123.
Cirri, 147.
Cetacea, 205.
Chiton, 226.
C. officinalis, 235.
Ciliograde *acalephæ*, 209.
Common Sea cucumber, 305.

Doris, 223, 363.
Doris (Spawn of), 226.
Diminutive Sucker-fish, 350.
Decapoda (ten-footed crustacea), 69, 85, 115.
D. sanguinea, 78, 118, 368.
Dorsibranchiate annelidans, 155.
Devil's hand, 279.
Dulse, 310.

Exuviation of Crabs, &c., 85, 113, 120, 132.
Exuviation of Prawns, &c., 139.
Exuviation of Barnacles, 147.

INDEX.

Eolis, 223.
Eolis, (Spawn of) 228.
   papillosa, 228.
Echinus, 291.
   sphæra, 297.
Entomology, 23.
Egg Urchin, 257.

Foraminifera, 24, 30.
Fan-amphitrite, 161, 163.
Fishes 329.
Five-fingers (Star-fish), 291.

Gulf stream, 206.
Girdle of Venus, 209.
Green-pea urchin, 299.
Gunnel-fish, 346.
Goby (one-spotted), 309, 348.
Goby (two-spotted), 348.
Gasterosteus aculeatus, 352.
Golden willow, 87.

Hyas araneus, 80.
Hermit crabs, 94, 105, 108, 130.
Hydra tuba, 213.
Hydra gelatinosa, 213.
Hyaline stylet, 239.
Holothuriadæ, 303.

Infusoria, 33, 43, 227.
Iridea edulis, 74, 236.
Irish Moss, 101.

Jelly fish, 203.

Kerona silurus, 35.

Lepas anatifera (Ship barnacle), 148, 150.
Laminated nereis, 155.
Luidia fragillissima, 287.
Lepus marinus, 309.
Lettuce Ulva, 180, 364.
Limpet, 81, 98.
Lobster crabs, 94.
Lily stars, 271.
Lobster (The), 131, 290.
L. digitata (Oar weed), 277.
Lingthorn, 285.
Lump sucker, 347.
Maia squinado (Spider crab), 79, 80, 82, 167.
Mussel (*Mytilus edulis*), 82, 122, 167, 363.
Medusæ, 203, 210, 216.

Modera formosa, 217.
Montague Sucker-fish, 349.
Marine Aquariæ, 357.
Macroura (as Lobsters), &c.

Nereis, 155, 158.
Nudibranchiate gasteropoda, 226.

Ostrea edulis (oyster), 81.
Ophiuræ, 277.
      texturata, 277.
      albida, 277.
Ophiocoma rosula, 279.

Paramecium caudatum, 35.
Portunus puber (Velvet crab), 68.
Porcellana longicornis (Minute Porcelain crab), 79.
Periwinkle, 95.
Pagurus bernardhus (Hermit crab), 96, 108, 146.
Pecten, 108, 370.
Prawn, Common (*Palæmon squilla*), 137
Phyllodoce laminosa, 155.
Purpura lapillus (Whelk), 103, 168.
Portuguese man-of war, 212
Pholas (*crispata*), 84, 153, 236, 245, 253, 365.
Pholas (*dactylus*), 260.
Pentacrinus Europæus, 275.
Psolus phantapus, 304.
Pike-fish, 339.
Pinnotheres pisum (Common Pea-crab), 82.
Pinna, 83.
Partane, The (Edible crab), 65, 124.
Pulmonigrade acalephæ, 211.
Physograde acalephæ, 211.
Parthenogenesis, 218.

Rotifera, or Wheel-bearers, 36.
Rosy Feather star, **276.**
Rosy Heart urchin, 300.
Rockling (Five-bearded), 347.
Razor-fish, 323.

Stickleback, 25, 108, 352.
Sertularia, 41.
Ship Barnacle, 145.
Sea-Mouse, 267.
Star fishes, 273.
Solaster papposa, 167, **286.**
      endeca, 286.
Sea Urchins, 289.

INDEX.

Silky Spined urchin, 299.
Sea Cucumbers, 303.
Sea Hares, 309.
Serpulæ, 315.
Sabellæ, 315, 319.
Solen siliqua, Razor fish, 324.
Soldier crabs, 92.
Shrimps, 137.
Saxicavæ, 248.
Sucker fishes, 348.
Shore crab, 65, 68, 72, 120, 235.
Swimming crab, 66.
Spider crab, 78,
Salex vitellina (golden willow), 87.
Silver-Willie (*T. zizziphanus*), 191.

Terebella figulus (the potter), 191, 195.
   littoralis, 197.
Trepang, 304.

Top-shell, 132.
Trochus, 146.
Tubiculous annelids, 162, 194.
Tanks, 355.

Uraster rubens, 167, 281.
Ulva latissima, 98, 180, 367, 368.
Univalves, 94.
Urchins (Sea), 298.

Vorticellæ, 35, 38.
Velvet Fiddler crab, 66, 74.

Whelk (*Purpura lapillus*), 168.

Zoothamnium spirale, 43.
Zooids, 219.
Zoophytes, 47, 49.

www.ingramcontent.com/pod-product-compliance
Lightning Source LLC
Chambersburg PA
CBHW032013220426
43664CB00006B/225